A SHORT COURSE ON BANACH SPACE THEORY

LONDON MATHEMATICAL SOCIETY STUDENT TEXTS

Managing editor: Professor J. W. Bruce,
Department of Mathematics, University of Hull, UK

A SHORT COURSE ON BANACH SPACE THEORY

N. L. CAROTHERS

Bowling Green State University

CAMBRIDGE
UNIVERSITY PRESS

CAMBRIDGE UNIVERSITY PRESS
Cambridge, New York, Melbourne, Madrid, Cape Town, Singapore,
São Paulo, Delhi, Dubai, Tokyo, Mexico City

Cambridge University Press
The Edinburgh Building, Cambridge CB2 8RU, UK

Published in the United States of America by Cambridge University Press, New York

www.cambridge.org
Information on this title: www.cambridge.org/9780521603720

First published 2004

A catalogue record for this publication is available from the British Library

Library of Congress Cataloguing in Publication data
Carothers, N. L., 1952–
A short course on Banach space theory / N. L. Carothers.
p.   cm.
Includes bibliographical references and index.
ISBN 0-521-84283-2 – ISBN 0-521-60372-2 (pbk.)
1. Banach spaces.   I. Title.
QA322.2.C37   2004
515′.732 – dc22   2004045627

ISBN 978-0-521-84283-9 Hardback
ISBN 978-0-521-60372-0 Paperback

Contents

Preface

These are notes for a graduate topics course offered on several occasions to a rather diverse group of doctoral students at Bowling Green State University. An earlier version of these notes was available through my Web pages for some time and, judging from the e-mail I've received, has found its way into the hands of more than a few readers around the world. Offering them in their current form seemed like the natural thing to do.

Although my primary purpose for the course was to train one or two students to begin doing research in Banach space theory, I felt obliged to present the material as a series of compartmentalized topics, at least some of which might appeal to the nonspecialist. I managed to cover enough topics to suit my purposes and, in the end, assembled a reasonable survey of at least the rudimentary tricks of the trade.

As a prerequisite, the students all had a two-semester course in real analysis that included abstract measure theory along with an introduction to functional analysis. While abstract measure theory is only truly needed in the final chapter, elementary facts from functional analysis, such as the Hahn–Banach theorem, the Open Mapping theorem, and so on, are needed throughout. Chapter 2, "Preliminaries," offers a brief summary of several key ideas from functional analysis, but it is far from self-contained. This chapter also features a large set of exercises I used as the basis for additional review, when necessary. A modest background in topology is also helpful but, because many of my students needed review here, I included a brief appendix containing most of the essential facts.

I make no claims of originality here. In fact, the presentation borrows heavily from several well-known sources. I tried my best to document these sources fully in the references and in the brief Notes and Remarks sections at the end of each chapter. You will also see that I've included a few exercises to accompany each chapter. These only scratch the surface, of course. Energetic readers may want to seek out greater challenges through the readings suggested in the Notes and Remarks.

My goal was a quick survey of what I perceive to be the major topics in classical Banach space theory: Basis theory, L_p spaces, $C(K)$ spaces, and a brief introduction to the geometry of Banach spaces. But the emphasis here is on *classical*; most of this material is more than thirty years old and, indeed, a great deal of it is more than fifty years old. Readers interested in contemporary research topics in Banach space theory are sure to be disappointed with this modest introduction and are encouraged to look elsewhere.

Finally, I should point out that the course has proven to be of interest to more students than I had originally imagined. Basis theory, for example, has enjoyed a resurgence in certain modern arenas, and such chestnuts as the so-called gliding hump argument frequently resurface in a variety of contemporary research venues. From this point of view, the course has much to offer students interested in operator theory, frames and wavelets, and even in certain corners of algebra such as lattice theory. More important, at least from my point of view, is that the early history of Banach space theory is loaded with elegant, insightful arguments and clever techniques that are not only worthy of study in their own right but are also deserving of greater publicity. It is in this spirit that I offer these notes.

Neal Carothers
Bowling Green, Ohio
February 2003

Chapter 1

Classical Banach Spaces

To begin, recall that a Banach space is a complete normed linear space. That is, a Banach space is a normed vector space $(X, \| \cdot \|)$ that is a complete metric space under the induced metric $d(x, y) = \|x - y\|$. Unless otherwise specified, we'll assume that all vector spaces are over $\mathbb{R}$, although, from time to time, we will have occasion to consider vector spaces over $\mathbb{C}$.

What follows is a list of the *classical* Banach spaces. Roughly translated, this means the spaces known to Banach. Once we have these examples out in the open, we'll have plenty of time to fill in any unexplained terminology. For now, just let the words wash over you.

The Sequence Spaces ℓ_p and c_0

Arguably the first infinite-dimensional Banach spaces to be studied were the sequence spaces ℓ_p and c_0. To consolidate notation, we first define the vector space s of all real sequences $x = (x_n)$ and then define various subspaces of s.

For each $1 \leq p < \infty$, we define

$$\|x\|_p = \left(\sum_{n=1}^{\infty} |x_n|^p \right)^{1/p}$$

and take ℓ_p to be the collection of those $x \in s$ for which $\|x\|_p < \infty$. The inequalities of Hölder and Minkowski show that ℓ_p is a normed space; from there it's not hard to see that ℓ_p is actually a Banach space.

The space ℓ_p is defined in exactly the same way for $0 < p < 1$ but, in this case, $\| \cdot \|_p$ defines a complete quasi-norm. That is, the triangle inequality now holds with an extra constant; specifically, $\|x + y\|_p \leq 2^{1/p}(\|x\|_p + \|y\|_y)$. It's worth noting that $d(x, y) = \|x - y\|_p^p$ defines a complete, translation-invariant metric on ℓ_p for $0 < p < 1$.

1

For $p = \infty$, we define ℓ_∞ to be the collection of all bounded sequences; that is, ℓ_∞ consists of those $x \in s$ for which

$$\|x\|_\infty = \sup_n |x_n| < \infty.$$

It's easy to see that convergence in ℓ_∞ is the same as uniform convergence on $\mathbb{N}$ and, hence, that ℓ_∞ is complete. There are two very natural (closed) subspaces of ℓ_∞: The space c, consisting of all convergent sequences, and the space c_0, consisting of all sequences converging to 0. It's not hard to see that c and c_0 are also Banach spaces.

As subsets of s we have

$$\ell_1 \subset \ell_p \subset \ell_q \subset c_0 \subset c \subset \ell_\infty \tag{1.1}$$

for any $1 < p < q < \infty$. Moreover, each of the inclusions is norm one:

$$\|x\|_1 \geq \|x\|_p \geq \|x\|_q \geq \|x\|_\infty. \tag{1.2}$$

It's of some interest here to point out that, although s is not itself a normed space, it is, at least, a complete metric space under the so-called Fréchet metric

$$d(x, y) = \sum_{n=1}^{\infty} 2^{-n} \frac{|x_n - y_n|}{1 + |x_n - y_n|}. \tag{1.3}$$

Clearly, convergence in the Fréchet metric implies coordinatewise convergence.

Finite-Dimensional Spaces

We will also have occasion to consider the finite-dimensional versions of the ℓ_p spaces. We write ℓ_p^n to denote $\mathbb{R}^n$ under the ℓ_p norm. That is, ℓ_p^n is the space of all sequences $x = (x_1, \ldots, x_n)$ of length n and is supplied with the norm

$$\|x\|_p = \left(\sum_{i=1}^{n} |x_i|^p \right)^{1/p}$$

for $p < \infty$, and

$$\|x\|_\infty = \max_{1 \leq i \leq n} |x_i|$$

for $p = \infty$.

Recall that all norms on $\mathbb{R}^n$ are equivalent. In particular, given any norm $\| \cdot \|$ on $\mathbb{R}^n$, we can find a positive, finite constant C such that

$$C^{-1} \|x\|_1 \leq \|x\| \leq C \|x\|_1 \tag{1.4}$$

for all $x = (x_1, \ldots, x_n)$ in $\mathbb{R}^n$. Thus, convergence in any norm on $\mathbb{R}^n$ is the same as "coordinatewise" convergence and, hence, every norm on $\mathbb{R}^n$ is complete.

Because every finite-dimensional normed space is just "$\mathbb{R}^n$ in disguise," it follows that every finite-dimensional normed space is complete.

The L_p Spaces

We first define the vector space $L_0[0, 1]$ to be the collection of all (equivalence classes, under equality almost everywhere [a.e.], of) Lebesgue-measurable functions $f : [0, 1] \to \mathbb{R}$. For our purposes, L_0 will serve as the "measurable analogue" of the sequence space s.

For $1 \le p < \infty$, the Banach space $L_p[0, 1]$ consists of those $f \in L_0[0, 1]$ for which

$$\|f\|_p = \left(\int_0^1 |f(x)|^p \, dx \right)^{1/p} < \infty.$$

The space $L_\infty[0, 1]$ consists of all (essentially) bounded $f \in L_0[0, 1]$ under the essential supremum norm

$$\|f\|_\infty = \operatorname*{ess.sup}_{0 \le x \le 1} |f(x)| = \inf \{ B : |f| \le B \text{ a.e.} \}$$

(in practice, though, we often just write "sup" in place of "ess.sup"). Again, the inequalities of Hölder and Minkowski play an important role here.

As before, the spaces $L_p[0, 1]$ are also defined for $0 < p < 1$, but $\| \cdot \|_p$ defines only a quasi-norm. Again, $d(f, g) = \|f - g\|_p^p$ defines a complete, translation-invariant metric on L_p for $0 < p < 1$. The space $L_0[0, 1]$ is given the topology of convergence (locally) in measure. For Lebesgue measure on $[0, 1]$, this topology is known to be equivalent to that given by the metric

$$d(f, g) = \int_0^1 \frac{|f(x) - g(x)|}{1 + |f(x) - g(x)|} \, dx. \tag{1.5}$$

As subsets of $L_0[0, 1]$, we have the following inclusions:

$$L_1[0, 1] \supset L_p[0, 1] \supset L_q[0, 1] \supset L_\infty[0, 1], \tag{1.6}$$

for any $1 < p < q < \infty$. Moreover, the inclusion maps are all norm one:

$$\|f\|_1 \le \|f\|_p \le \|f\|_q \le \|f\|_\infty. \tag{1.7}$$

The spaces $L_p(\mathbb{R})$ are defined in much the same way but satisfy *no* such inclusion relations. That is, for any $p \ne q$, we have $L_p(\mathbb{R}) \not\subset L_q(\mathbb{R})$. Nevertheless, you may find it curious to learn that $L_p(\mathbb{R})$ and $L_p[0, 1]$ are linearly *isometric*.

More generally, given a measure space (X, Σ, μ), we might consider the space $L_p(\mu)$ consisting of all (equivalence classes of) Σ-measurable functions $f : X \to \mathbb{R}$ under the norm

$$\|f\|_p = \left(\int_X |f(x)|^p \, d\mu(x) \right)^{1/p}$$

(with the obvious modification for $p = \infty$).

It is convenient to consider at least one special case here: Given any set Γ, we define $\ell_p(\Gamma) = L_p(\Gamma, 2^\Gamma, \mu)$, where μ is counting measure on Γ. What this means is that we identify functions $f : \Gamma \to \mathbb{R}$ with "sequences" $x = (x_\gamma)$ in the usual way: $x_\gamma = f(\gamma)$, and we define

$$\|x\|_p = \left(\sum_{\gamma \in \Gamma} |x_\gamma|^p \right)^{1/p} = \left(\int_\Gamma |f(\gamma)|^p \, d\mu(\gamma) \right)^{1/p} = \|f\|_p$$

for $p < \infty$. Please note that if $x \in \ell_p(\Gamma)$, then $x_\gamma = 0$ for all but countably many γ. For $p = \infty$, we set

$$\|x\|_\infty = \sup_{\gamma \in \Gamma} |x_\gamma| = \sup_{\gamma \in \Gamma} |f(\gamma)| = \|f\|_\infty.$$

We also define $c_0(\Gamma)$ to be the space of all those $x \in \ell_\infty(\Gamma)$ for which the set $\{\gamma : |x_\gamma| > \varepsilon\}$ is *finite* for any $\varepsilon > 0$. Again, this forces an element of $c_0(\Gamma)$ to have countable support. Clearly, $\ell_p(\mathbb{N}) = \ell_p$ and $c_0(\mathbb{N}) = c_0$.

A priori, the Banach space characteristics of $L_p(\mu)$ will depend on the underlying measure space (X, Σ, μ). As it happens, though, Lebesgue measure on $[0, 1]$ and counting measure on $\mathbb{N}$ are essentially the only two cases we have to worry about. It follows from a deep result in abstract measure theory (Maharam's theorem [97]) that every complete measure space can be decomposed into "nonatomic" parts (copies of $[0, 1]$) and "purely atomic" parts (counting measure on some discrete space). From a Banach space point of view, this means that every L_p space can be written as a direct sum of copies of $L_p[0, 1]$ and $\ell_p(\Gamma)$ (or ℓ_p^n).

For the most part we will divide our efforts here into three avenues of attack: Those properties of L_p spaces that don't depend on the underlying measure space, those that are peculiar to $L_p[0, 1]$, and those that are peculiar to the ℓ_p spaces.

The $C(K)$ Spaces

Perhaps the earliest known example of a Banach space is the space $C[a, b]$ of all continuous real-valued functions $f : [a, b] \to \mathbb{R}$ supplied with the "uniform norm":

$$\|f\| = \max_{a \le t \le b} |f(t)|.$$

More generally, if K is any compact Hausdorff space, we write $C(K)$ to denote the Banach space of all continuous real-valued functions $f : K \to \mathbb{R}$ under the norm

$$\|f\| = \max_{t \in K} |f(t)|.$$

For obvious reasons, we sometimes write the norm in $C(K)$ as $\|f\|_\infty$ and refer to it as the "sup norm." In any case, convergence in $C(K)$ is the same as uniform convergence on K.

In Banach's day, point set topology was still very much in its developmental stages. In his book [6], Banach considered $C(K)$ spaces only in the case of compact *metric* spaces K. We, on the other hand, may have occasion to venture further. At the very least, we will consider the case in which K is a compact Hausdorff space (since the theory is nearly identical in this case). And, if we really get ambitious, we may delve into more esoteric settings. For the sake of future reference, here is a brief summary of the situation.

If X is any topological space, we write $C(X)$ to denote the algebra of all real-valued continuous functions $f : X \to \mathbb{R}$. For general X, though, $C(X)$ may not be metrizable. If X is Hausdorff and σ-compact, say $X = \bigcup_{n=1}^\infty K_n$, then $C(X)$ is a complete metric space under the topology of "uniform convergence on compacta" (or the "compact-open" topology). This topology is generated by the so-called Fréchet metric

$$d(f, g) = \sum_{n=1}^\infty 2^{-n} \frac{\|f - g\|_n}{1 + \|f - g\|_n}, \tag{1.8}$$

where $\|f\|_n$ is the norm of $f|_{K_n}$ in $C(K_n)$.

If we restrict our attention to the *bounded* functions in $C(X)$, then we may at least apply the sup norm; for this reason, we consider instead the Banach space $C_b(X)$ of all bounded, continuous, real-valued functions $f : X \to \mathbb{R}$ endowed with the sup norm

$$\|f\| = \sup_{x \in X} |f(x)|.$$

Obviously, $C_b(X)$ is a closed subspace of $\ell_\infty(X)$. If X is at least completely regular, then $C_b(X)$ contains as much information as $C(X)$ itself in the sense that the topology on X is completely determined by knowing the bounded, continuous, real-valued functions on X.

If X is noncompact, then we might also consider the normed space $C_C(X)$ of all continuous $f : X \to \mathbb{R}$ with *compact support*. That is, $f \in C_C(X)$ if f is continuous and if the *support* of f, namely, the set

$$\mathrm{supp}\, f = \overline{\{x \in X : f(x) \neq 0\}},$$

is compact. Although we may apply the sup norm to $C_C(X)$, it's not, in general, complete. The completion of $C_C(X)$ is the space $C_0(X)$ consisting of all those continuous $f : X \to \mathbb{R}$ that "vanish at infinity." Specifically, $f \in C_0(X)$ if f is continuous and if, for each $\varepsilon > 0$, the set $\{|f| \geq \varepsilon\}$ has compact closure. The space $C_0(X)$ is a closed subspace of $C_b(X)$ and hence is a Banach space under the sup norm.

If X is compact, then, of course, $C_C(X) = C_b(X) = C(X)$. For general X, however, the best we can say is

$$C_C(X) \subset C_0(X) \subset C_b(X) \subset C(X).$$

At least one easy example might be enlightening here: Consider the case $X = \mathbb{N}$; obviously, $\mathbb{N}$ is locally compact and metrizable. Now *every* function $f : \mathbb{N} \to \mathbb{R}$ is continuous, and any such function can quite plainly be identified with a sequence; namely, its range $(f(n))$. That is, we can identify $C(\mathbb{N})$ with s by way of the correspondence $f \in C(\mathbb{N}) \longleftrightarrow x \in s$, where $x_n = f(n)$. Convince yourself that

$$C_b(\mathbb{N}) = \ell_\infty, \quad C_0(\mathbb{N}) = c_0, \quad C_0(\mathbb{N}) \oplus \mathbb{R} = c, \qquad (1.9)$$

and that

$$C_C(\mathbb{N}) = \{x \in s : x_n = 0 \ \text{for all but finitely many } n\}. \qquad (1.10)$$

While this is curious, it doesn't quite tell the whole story. Indeed, both ℓ_∞ and c are actually $C(K)$ spaces. To get a glimpse into why this is true, consider the space $\mathbb{N}^* = \mathbb{N} \cup \{\infty\}$, the *one-point compactification* of $\mathbb{N}$ (that is, we append a "point at infinity"). If we define a neighborhood of ∞ to be any set with finite (compact) complement, then $\mathbb{N}^*$ becomes a compact Hausdorff space. Convince yourself that

$$c = C(\mathbb{N}^*) \qquad \text{and} \qquad c_0 = \{f \in C(\mathbb{N}^*) : f(\infty) = 0\}. \qquad (1.11)$$

We'll have more to say about these ideas later.

Hilbert Space

As you'll no doubt recall, the spaces ℓ_2 and L_2 are both Hilbert spaces, or complete inner product spaces. Recall that a vector space H is called a Hilbert space if H is endowed with an inner product $\langle \cdot, \cdot \rangle$ with the property that the induced norm, defined by

$$\|x\| = \sqrt{\langle x, x \rangle},\qquad (1.12)$$

is complete. It is most important here to recognize that the norm in H is intimately related to an inner product by way of (1.12). This is a tall order for the run-of-the-mill norm. From this point of view, Hilbert spaces are quite rare among the teeming masses of Banach spaces.

There is a critical distinction to be made here; perhaps an example will help to explain. Let X denote the space ℓ_2 supplied with the norm $\|x\| = \|x\|_2 + \|x\|_\infty$. Then X is isomorphic (linearly homeomorphic) to ℓ_2 because our new norm satisfies $\|x\|_2 \le \|x\| \le 2\|x\|_2$. But X is *not* itself a Hilbert space. The test is whether the *parallelogram law* holds:

$$\|x + y\|^2 + \|x - y\|^2 \overset{?}{=} 2 \left(\|x\|^2 + \|y\|^2 \right).$$

And it's easy to check that the parallelogram law fails if $x = (1, 0, 0, \ldots)$ and $y = (0, 1, 0, \ldots)$, for instance. The moral here is that it's not enough to have a well-defined inner product, nor is it enough to have a norm that is close to a known Hilbert space norm. In a Hilbert space, the norm and the inner product are inextricably bound together through equation (1.12).

Hilbert spaces exhibit another property that is rare among the Banach spaces: In a Hilbert space, every closed subspace is the range of a continuous projection. This is far from the case in a general Banach space. (In fact, it is known that any space with this property is already isomorphic to Hilbert space.)

"Neoclassical" Spaces

We have more or less exhausted the list of spaces that were well known in Banach's time. But we have by no means even begun to list the spaces that are commonplace these days. In fact, it would take pages and pages of definitions to do so. For now we'll content ourselves with the understanding that all of the known examples are, in a sense, generalizations of the spaces we have seen thus far.

The Big Questions

We're typically interested in both the *isometric* as well as the *isomorphic* character of a Banach space. (For our purposes, all isometries are *linear.*

Also, as a reminder, an isomorphism is a *linear* homeomorphism.) Here are just a few of the questions we might consider:

▷ Are all the spaces listed above isometrically distinct? For example, is it at all possible that ℓ_4 and ℓ_6 are isometric? What about ℓ_p and L_p? Or $L_p[0, 1]$ and $L_p(\mathbb{R})$?

▷ When is a given Banach space X isometric to a subspace of one of the classical spaces? When does X contain an isometric copy of one of the classical spaces? In particular, does L_1 embed isometrically into L_2? Does ℓ_p embed isometrically into $C[0, 1]$?

▷ We might pose all of the preceding questions, replacing the word "isometric" with "isomorphic."

▷ Characterize all of the subspaces of a given Banach space X, if possible, both isometrically and isomorphically. In particular, identify those subspaces that are the range of a continuous projection (that is, the *complemented* subspaces of X). Knowing all of the subspaces of a given space would tell us something about the linear operators into or on the space. (And vice versa: After all, the kernel and range of a linear operator are subspaces.)

▷ All of the spaces we've defined above carry some additional structure. $C[a, b]$ is an algebra of functions, for example, and $L_1[0, 1]$ is a lattice. What, if anything, does this extra structure tell us from the point of view of Banach space theory? Is it an isometric invariant of these spaces? An isomorphic invariant? Does it imply the existence of interesting subspaces? Or interesting operators?

▷ It's probably fair to say that functional analysis concerns the study of *operators* between spaces. Insert the adjective "linear," wherever possible, and you will have a working definition of *linear* functional analysis. Where does the study of Banach spaces fit within the larger field of functional analysis? In other words, does a better understanding of Banach spaces tell us anything about the operators between these spaces?

▷ Good mathematics doesn't exist in a vacuum. We also want to keep an eye out for applications of the theory of Banach spaces to "mainstream" analysis. Conversely, we will want to be on the lookout for applications of mainstream tools to the theory of Banach spaces. Among others, we will look for connections with probability, harmonic analysis, topology, operator theory, and plain ol' calculus. By way of an example, we might consider the calculus of "vector-valued" functions $f : [0, 1] \rightarrow X$, where X is a Banach space. It would make perfect sense to ask whether f is of bounded variation, for instance, or whether $\int_0^1 \|f(x)\|\,dx < \infty$. We'll put these tantalizing questions aside until we're better prepared to deal with them.

Notes and Remarks

The space $C[a, b]$ is arguably the oldest of the examples presented here; it was Maurice Fréchet who offered the first systematic study of the space (as a metric space) beginning in 1906. The space ℓ_2 was introduced in 1907 by Erhard Schmidt (of the "Gram–Schmidt process"). The space that bears his name held little interest for Hilbert, by the way. Hilbert preferred the concrete setting of integral equations to the abstractions of Hilbert space theory.

Schmidt's paper is notable in that it is believed to contain the first appearance of the familiar "double-bar" notation for norms. Both the notation ℓ_2 and the attribution "Hilbert space," though, are due to Frigyes (Frederic) Riesz. In fact, Riesz introduced the L_p spaces, and he, Fréchet, and Ernst Fischer noticed their connections with the ℓ_p spaces. Although many of these ideas were "in the air" for several years, it was Banach who launched the first *comprehensive* study of normed spaces in his 1922 dissertation [5], culminating in his 1932 book [6]. For more on the prehistory of functional analysis and, in particular, the development of function spaces, see the detailed articles by Michael Bernkopf [13, 14], the writings of A. F. Monna [104, 105], and the excellent chapter notes in Dunford and Schwartz [42].

For much more on the classical and "neoclassical" Banach spaces, see the books by Adams [1], Bennett and Sharpley [12], Dunford and Schwartz [42], Duren [43], Lacey [88], Lindenstrauss and Tzafriri [93, 94, 95], and Wojtaszczyk [147]. For more on the history of open questions and unresolved issues in Banach space theory, see Banach's book [6], its review by Diestel [32], and its English translation with notes by Bessaga and Pełczyński [7]; see also Day [29], Diestel [33], Diestel and Uhl [34], Megginson [100], and the articles by Casazza [20, 21, 22, 23], Mascioni [99], and Rosenthal [124, 125, 126, 127].

Exercises

1. If $(X, \|\cdot\|)$ is any normed linear space, show that the operations $(x, y) \mapsto x + y$ and $(\alpha, x) \mapsto \alpha x$ are continuous (on $X \times X$ and $\mathbb{R} \times X$, respectively). [It doesn't much matter what norms we use on $X \times X$ and $\mathbb{R} \times X$; for example, $\|(x, y)\| = \|x\| + \|y\|$ works just fine. (Why?)] If Y is a (linear) subspace of X, conclude that its closure $\overline{Y}$ is again a subspace.

2. Show that $(X, \| \cdot \|)$ is complete if and only if every absolutely summable series is summable; that is, if and only if $\sum_{n=1}^{\infty} \|x_n\| < \infty$ always implies that $\sum_{n=1}^{\infty} x_n$ converges in (the norm of) X.

3. Show that $C^{(1)}[0, 1]$, the space of functions $f : [0, 1] \to \mathbb{R}$ having a continuous first derivative, is complete under the norm $\|f\| = \|f\|_\infty + \|f'\|_\infty$.

4. Show that s is complete under the Fréchet metric (1.3).

5. Show that ℓ_∞ is not separable.

6. Given $0 < p < 1$, show that $\|f + g\|_p \le 2^{1/p}(\|f\|_p + \|g\|_p)$ for $f, g \in L_p$. A better estimate (with a slightly harder proof) yields the constant $2^{(1/p)-1}$ in place of $2^{1/p}$.

7. Let $1 < p < \infty$ and let $1/p + 1/q = 1$. Show that for positive real numbers a and b we have $ab \le a^p/p + b^q/q$ with equality if and only if $a = b$. For $0 < p < 1$ (and $q < 0$!), show that the inequality reverses.

8. Let $0 < p < 1$ and let $1/p + 1/q = 1$. If f and g are nonnnegative functions with $f \in L_p$ and $\int g^q > 0$, show that $\int fg \ge (\int f^p)^{1/p}(\int g^q)^{1/q}$.

9. Given $0 < p < 1$ and nonnegative functions $f, g \in L_p$, show that $\|f + g\|_p \ge \|f\|_p + \|g\|_p$.

10. Prove the string of inequalities (1.2) for $x \in \ell_1$.

11. Prove the string of inequalities (1.7) for $f \in L_\infty[0, 1]$.

12. Given $1 \le p, q \le \infty$, $p \ne q$, show that $L_p(\mathbb{R}) \not\subset L_q(\mathbb{R})$.

13. Given a compact Hausdorff space X, show that $C_0(X)$ is a closed subspace of $C_b(X)$ and that $C_C(X)$ is dense in $C_0(X)$.

14. Let H be a separable Hilbert space with orthonormal basis (e_n) and let K be a compact subset of H. Given $\varepsilon > 0$, show there exists an N such that $\| \sum_{n=N}^{\infty} \langle x, e_n \rangle e_n \| < \varepsilon$ for every $x \in K$. That is, if K is compact, then these "tail series" can be made *uniformly* small over K.

Chapter 2

Preliminaries

We begin with a brief summary of important facts from functional analysis – some with proofs, some without. Throughout, X, Y, and so on, are normed linear spaces over $\mathbb{R}$. If there is no danger of confusion, we will use $\| \cdot \|$ to denote the norm in any given normed space; if two or more spaces enter into the discussion, we will use $\| \cdot \|_X$, and so forth, to further identify the norm in question.

Continuous Linear Operators

Given a linear map $T : X \to Y$, recall that the following are equivalent:

(i) T is continuous at $0 \in X$.
(ii) T is continuous.
(iii) T is uniformly continuous.
(iv) T is Lipschitz; that is, there exists a constant $C < \infty$ such that $\|Tx - Ty\|_Y \leq C\|x - y\|_X$ for all $x, y \in X$.
(v) T is *bounded*; that is, there exists a constant $C < \infty$ such that $\|Tx\|_Y \leq C\|x\|_X$ for all $x \in X$.

If a linear map $T : X \to Y$ is bounded, then there is, in fact, a *smallest* constant C satisfying $\|Tx\|_Y \leq C\|x\|_X$ for all $x \in X$. Indeed, the constant

$$\|T\| = \sup_{x \neq 0} \frac{\|Tx\|_Y}{\|x\|_X} = \sup_{\|x\|_X \leq 1} \|Tx\|_Y, \tag{2.1}$$

called the *norm* of T, works; that is, it satisfies $\|Tx\|_Y \leq \|T\|\|x\|_X$ and it's the smallest constant to do so. Further, it's not hard to see that (2.1) actually defines a norm on the space $B(X, Y)$ of all bounded, continuous, linear maps $T : X \to Y$.

A map $T : X \to Y$ is called an *isometry* (into) if $\|Tx - Ty\|_Y = \|x - y\|_X$ for all $x, y \in X$. Clearly, every isometry is continuous. If, in addition, T is linear, then it's only necessary to check that $\|Tx\|_Y = \|x\|_X$ for every $x \in X$.

11

Throughout these notes, unless otherwise specified, the word "isometry" will always mean "linear isometry." (It's a curious fact, due to Banach and Mazur [8, 6], that any isometry mapping 0 to 0 is actually linear.) Please note that an onto isometry is invertible and the inverse map is again an isometry.

A linear map T from X onto Y is called an *isomorphism* if T is one-to-one and both T and T^{-1} are continuous. That is, T is an isomorphism (onto) if T is a linear homeomorphism. It follows from our first observation that T is an isomorphism if and only if T is onto and there exist constants $0 < c, C < \infty$ such that $c\|x\|_X \le \|Tx\|_Y \le C\|x\|_X$ for all $x \in X$. If we drop the requirement that T be onto, then this pair of inequalities defines an *into isomorphism* (that is, an isomorphism onto the range of T). Note that if X is a Banach space, then every isomorph of X is necessarily also a Banach space. It follows from the Open Mapping theorem that a one-to-one, onto linear map T defined on a Banach space X is necessarily an isomorphism.

Phrases such as "X is isometric to Y" or "Y contains an isomorphic copy of X" should be self-explanatory.

Finite-Dimensional Spaces

Every finite-dimensional normed space is just $\mathbb{R}^n$ in disguise. To see this, we first check that all norms on $\mathbb{R}^n$ are equivalent. To this end, let $\|\cdot\|$ be any norm on $\mathbb{R}^n$. We will find constants $0 < c, C < \infty$ such that

$$c\|x\|_1 \le \|x\| \le C\|x\|_1 \qquad (2.2)$$

for all $x \in \mathbb{R}^n$.

Let $e_1, \ldots, e_n$ be the usual basis for $\mathbb{R}^n$. Then, given $x = \sum_{i=1}^n a_i e_i \in \mathbb{R}^n$, we have

$$\|x\| = \left\|\sum_{i=1}^n a_i e_i\right\| \le \sum_{i=1}^n |a_i|\|e_i\| \le \left(\max_{1\le i\le n}\|e_i\|\right)\sum_{i=1}^n |a_i| = C\|x\|_1,$$

where $C = \max_{1\le i\le n}\|e_i\|$. The hard work comes in establishing the other inequality.

Now the inequality that we've just established shows that the function $\|\cdot\|$ is continuous on $(\mathbb{R}^n, \|\cdot\|_1)$. Indeed,

$$\left|\|x\| - \|y\|\right| \le \|x - y\| \le C\|x - y\|_1.$$

Hence, its restriction to $S = \{x : \|x\|_1 = 1\}$ is likewise continuous. And, since S is compact, $\|\cdot\|$ must attain a minimum value on S. What this means is that there exists some constant c such that $\|x\| \ge c$ whenever $\|x\|_1 = 1$. Thus, for any $x \in \mathbb{R}^n$, we have $\|x\| \ge c\|x\|_1$ by homogeneity. Since we may assume that the value c is actually attained, we must have $c > 0$, and so we're done.

Next, suppose that X is any finite-dimensional normed space with basis $x_1, \ldots, x_n$. Then the basis-to-basis map $x_i \mapsto e_i$ extends to a linear isomorphism from X onto $\mathbb{R}^n$. Indeed, if we define a new norm on $\mathbb{R}^n$ by setting

$$\left\| \sum_{i=1}^{n} a_i e_i \right\| = \left\| \sum_{i=1}^{n} a_i x_i \right\|_X,$$

then $\| \cdot \|$ must be equivalent to the usual norm on $\mathbb{R}^n$. Hence,

$$c \left\| \sum_{i=1}^{n} a_i x_i \right\|_X \leq \left\| \sum_{i=1}^{n} a_i e_i \right\|_2 \leq C \left\| \sum_{i=1}^{n} a_i x_i \right\|_X$$

for some constants $0 < c, C < \infty$.

As an immediate corollary, we obtain that every finite-dimensional normed space is complete. And, in particular, if X is a finite-dimensional subspace of a normed space Y, then X must be *closed* in Y.

Related to this is the fact that a normed space X is finite dimensional if and only if $B_X = \{x : \|x\| \leq 1\}$, the closed unit ball in X, is *compact*. The forward implication is obvious. For the backward implication, we appeal to *Riesz's lemma*: For each closed subspace Y of X and each $0 < \varepsilon < 1$, there exists a norm one vector $x = x_\varepsilon \in X$ such that $\|x - y\| \geq 1 - \varepsilon$ for all $y \in Y$. Thus, if X is infinite dimensional, we can inductively construct a sequence of norm one vectors (x_n) in X such that $\|x_n - x_m\| \geq 1/2$ for all $n \neq m$.

Perhaps not so immediate is that *every linear map on a finite-dimensional space is continuous*. Indeed, suppose that X is finite dimensional and that $x_1, \ldots, x_n$ is a basis for X. Then, from (2.2) and our previous discussion,

$$\sum_{i=1}^{n} |a_i| \leq C \left\| \sum_{i=1}^{n} a_i x_i \right\|_X \tag{2.3}$$

for some constant $C < \infty$. Thus, if $T : X \to Y$ is linear, we have

$$\left\| T \left(\sum_{i=1}^{n} a_i x_i \right) \right\|_Y \leq \sum_{i=1}^{n} |a_i| \|T x_i\|_Y \leq C \left(\max_{1 \leq i \leq n} \|T x_i\|_Y \right) \left\| \sum_{i=1}^{n} a_i x_i \right\|_X.$$

That is, $\|T x\|_Y \leq K \|x\|_X$, where $K = C \cdot \max_{1 \leq i \leq n} \|T x_i\|_Y$.

Continuous Linear Functionals

A scalar-valued map $f : X \to \mathbb{R}$ is called a *functional*. From our earlier observations, a *linear* functional is continuous if and only if it is bounded;

that is, if and only if there is a constant $C < \infty$ such that $|f(x)| \leq C\|x\|$ for every $x \in X$. If f is bounded, then the smallest constant C that will work in this inequality is defined to be the *norm* of f, again written $\|f\|$. Specifically,

$$\|f\| = \sup_{x \neq 0} \frac{|f(x)|}{\|x\|} = \sup_{\|x\| \leq 1} |f(x)|.$$

In particular, note that $|f(x)| \leq \|f\|\|x\|$ for any $f \in X^*, x \in X$.

The collection X^* of continuous, linear functionals on X can be identified with the collection of bounded (in the ordinary sense), continuous, linear (in a restricted sense), real-valued functions on B_X, the closed unit ball of X. Indeed, it's easy to check that, under this identification, X^* is a closed subspace of $C_b(B_X)$. In particular, X^* *is always a Banach space*. In language borrowed from linear algebra, X^* is called the *dual space* to X or, sometimes, the *norm* dual, to distinguish it from the *algebraic* dual (the collection of all linear functionals, continuous or not).

The word "dual" may not be the best choice here, for it is typically not true that the second dual space $X^{**} = (X^*)^*$ can be identified with X. If, for example, X is not complete, then there is little hope of X being isometrically identical to the complete space X^{**}. Nevertheless, X *is always isometric to a subspace of* X^{**}. This is a very important observation and is worth discussing in some detail.

Each element $x \in X$ induces a linear functional $\widehat{x}$ on X^* by way of point evaluation: $\widehat{x}(f) = f(x)$ for $f \in X^*$. Using the "inner product" notation $f(x) = \langle x, f \rangle$ to denote the action of a functional f on a vector x, note that the functional $\widehat{x}$ satisfies

$$\langle f, \widehat{x} \rangle = \langle x, f \rangle.$$

It's clear that $\widehat{x}$ is linear in f. Continuity is almost as easy: $|\widehat{x}(f)| = |f(x)| \leq \|x\|\|f\|$. Thus, $\widehat{x} \in X^{**}$ and $\|\widehat{x}\| \leq \|x\|$. But, from the Hahn–Banach theorem, we actually have $\|\widehat{x}\| = \|x\|$. Indeed, given any $0 \neq x \in X$, the Hahn–Banach theorem supplies a *norming functional*: A norm one element $f \in X^*$ such that $|f(x)| = \|x\|$. All that remains is to note that the correspondence $x \mapsto \widehat{x}$ is also linear. Thus, the map $x \mapsto \widehat{x}$ is actually a linear isometry from X into X^{**}.

We will also have occasion to write the "hat map" in more traditional terms using the operator $i : X \to X^{**}$ defined by $(i(x))(f) = f(x)$. In terms of the "inner product" notation, then, $i(x)$ satisfies

$$\langle f, i(x) \rangle = \langle x, f \rangle.$$

We will write $\widehat{X}$ to denote the image of X in X^{**} under the canonical "hat" embedding. As a closed subspace of a complete space, it's clear that $\overline{\widehat{X}}$, the closure of $\widehat{X}$ in X^{**}, is again complete. It follows that $\overline{\widehat{X}}$ is a *completion* of X. This is an important observation: For one, we now know that every normed space has a completion *that is again a normed space* and, for another, because X is always (isometric to) a subspace of a complete space, there's rarely any harm in simply assuming that X itself is complete. (Note, for example, that X and its completion necessarily have the same dual space.)

If it should happen that $\widehat{X} = X^{**}$, we say that X is *reflexive*. It's important to note here that this requirement is much more stringent than simply asking that X and X^{**} be isometric. Indeed, there is a famous example, due to R. C. James [73], of a Banach space J with the property that J and J^{**} are isometrically isomorphic (by way of an "unnatural" map) and yet $\widehat{J}$ is a *proper* closed subspace of J^{**}.

Adjoints

Each continuous, linear map $T : X \rightarrow Y$ induces a continuous, linear map $T^* : Y^* \rightarrow X^*$ called the *adjoint* of T. Indeed, given $f \in Y^*$, the composition $f \circ T$ defines an element of X^*. We define $T^*(f) = f \circ T$. Note that composition with T is linear. That T^* is continuous is easy: Clearly, $\|T^* f\| \leq \|T\| \|f\|$ and, hence, $\|T^*\| \leq \|T\|$. To see that $\|T^*\| = \|T\|$ actually holds, first choose a norm one vector $x \in X$ such that $\|Tx\| \geq \|T\| - \varepsilon$ and then choose a norm on functional $f \in Y^*$ such that $(T^* f)(x) = f(Tx) = \|Tx\|$. Then,

$$\|T^*\| \geq \|T^* f\| \geq (T^* f)(x) = \|Tx\| \geq \|T\| - \varepsilon.$$

Equivalently, T^* is defined by the formula

$$\langle x, T^* f \rangle = \langle Tx, f \rangle$$

for every $x \in X$ and $f \in Y^*$. In this notation, it's easy to see that T^* reduces to the familiar (conjugate) transpose of T in the case of matrix operators between $\mathbb{R}^n$ and $\mathbb{R}^m$.

Of course, it also makes sense to consider $T^{**} : X^{**} \rightarrow Y^{**}$, where $T^{**} = (T^*)^*$. Convince yourself that T^{**} is an extension of T, that is, under the usual convention that $X \subset X^{**}$ and $Y \subset Y^{**}$, we have $T^{**}|_X = T$. That $\|T^{**}\| = \|T\|$ follows from our previous remarks.

We will make occasional use of the following result; for a proof, see [100, Theorem 3.1.22], [26, Theorem 1.10], or [129, Theorem 4.15].

Theorem 2.1. *Let $T : X \rightarrow Y$ be a bounded linear map between Banach spaces X and Y. Then,*

(i) *T is onto if and only if T^* is an isomorphism into.*
(ii) *T is an isomorphism into if and only if T^* is onto.*

Projections

Given subspaces M and N of a vector space X, we write $X = M \oplus N$ if it's the case that each $x \in X$ can be uniquely written as $x = y + z$ with $y \in M$ and $z \in N$. In this case, we say that X is the *direct sum* of M and N or that M and N are *complements* in X. Given a single subspace M of X, we say that M is *complemented* in X if we can find a complementary subspace N; that is, if we can write $X = M \oplus N$ for some N. It's easy to see that complements need not be unique.

If $X = M \oplus N$, then we can define a map $P : X \rightarrow X$ by $Px = y$ where $x = y + z$, as above. Uniqueness of the splitting $x = y + z$ shows that P is well-defined and *linear*. Clearly, the range of P is M; in fact, $P^2 = P$; hence, $P|_M$ is the identity on M. Equally clear is that N is the kernel of P. In short, P is a linear *projection* (an idempotent: $P^2 = P$) on X with range M and kernel N.

Conversely, given a linear projection $P : X \rightarrow X$, it's easy to see that we can write $X = M \oplus N$, where M is the range of P and N is the kernel of P. Indeed, $x = Px + (I - P)x$, and $P(I - P)x = Px - P^2x = 0$ for any $x \in X$. While we're at it, notice that $Q = I - P$ is also a projection (the range of Q is the kernel of P; the kernel of Q is the range of P). As before, though, typically there are many different projections on X with a given range (or kernel).

In summary, given a subspace M of X, finding a complement for M in X is equivalent to finding a linear projection P on X with range M. But we're interested in *normed* vector spaces and *continuous* maps. If X is a normed linear space, when is a subspace M the range of a *continuous* linear projection? Well, a moment's reflection will convince you that the range of a continuous projection must be *closed*. Indeed, if $Px_n \rightarrow y$, then, by continuity, $P^2x_n = Px_n \rightarrow Py$. Thus, we must have $Py = y$; that is, y must be in the range of P. Of course, if P is continuous, then $N = \ker P$ is also closed.

Conversely, if M is closed, and if we can write $X = M \oplus N$ for some *closed* subspace N of X, then the corresponding projection P with kernel N is necessarily continuous. This follows from an easy application of the Closed

Graph theorem: Suppose that $x_n \to x$ and $Px_n \to y$. Since $Px_n \in M$, and since M is closed, we must have $y \in M$, too. Now we need to show that $y = Px$, and for this it suffices to show that $x - y \in N$. But $x_n - Px_n = (I - P)x_n \to x - y$ and $(I - P)x_n \in N$ for each N. Thus, since N is also closed, we must have $x - y \in N$. (Please note that we needed *both* M and N to be closed for this argument.)

Henceforth, given a *closed* subspace M of a normed linear space X, we will say that M is *complemented* in X if there is another *closed* subspace N such that $X = M \oplus N$. Equivalently, M is complemented in X if M is the range of a *continuous* linear projection P on X. It will take some time before we can fill in the details, but you may find it enlightening to hear that there do exist closed, uncomplemented subspaces of certain Banach spaces. In fact, outside of the Hilbert space setting, nontrivial projections are often hard to come by.

Example. If M is a one-dimensional subspace of a normed space X, then M is complemented in X.

Proof. Given $0 \neq y \in M$, the Hahn–Banach theorem provides a continuous linear functional $f \in X^*$ with $f(y) = 1$. Check that $Px = f(x)y$ is a continuous linear projection on X with range M. $\square$

Quotients

Given a subspace M of a vector space X, we next consider the *quotient space* X/M. The quotient space consists of all cosets, or equivalence classes, $[x] = x + M$, where $x \in X$. Two cosets $x + M$ and $y + M$ are declared equal if $x - y \in M$; that is, x and y are proclaimed equivalent if $x - y \in M$, and so we identify $[x]$ and $[y]$ in this case, too. Addition and scalar multiplication are defined in the obvious way: $[x] + [y] = [x + y]$, and $a[x] = [ax]$. It's easy to check that these operations make X/M a vector space (with zero vector $[0] = M$). We also define the *quotient map* $q : X \to X/M$ by $q(x) = [x]$. Under the operations we've defined on X/M, it's clear that q is *linear* (and necessarily onto) with kernel M.

Again, our interest lies in the case in which X is a normed space. In this case, we want to know whether there is a natural way to define a norm on X/M and whether this norm will make the quotient map q continuous. We take the easy way out here. We know that we want q to be continuous, and so we will force the norm on X/M to satisfy the inequality $\|q(x)\|_{X/M} \leq \|x\|_X$ for all $x \in X$. But $q(x) = q(x + y)$ for any $y \in M$; thus, we actually need the norm on X/M to satisfy $\|q(x)\|_{X/M} \leq \|x + y\|_X$ for all $x \in X$ and all $y \in M$. This leads us to define the *quotient norm* on X/M by

$$\|q(x)\|_{X/M} = \inf_{y \in M} \|x + y\|_X. \tag{2.4}$$

That is, $\|q(x)\|_{X/M} = d(x, M)$, the *distance* from x to M. Given this, we evidently have $\|q(x)\|_{X/M} = 0$ precisely when $x \in \overline{M}$. Thus, there is little hope of the quotient "norm" defining anything more than a pseudonorm unless we also insist that M be a *closed* subspace of X. And that's just what we'll do.

Now most of what we need to check follows easily from the fact that M is a subspace. For example, since $0 \in M$, we clearly have $\|q(x)\|_{X/M} \leq \|x\|_X$. Next,

$$\begin{aligned} \|aq(x)\|_{X/M} &= \|q(ax)\|_{X/M} \\ &= \inf_{y \in M} \|ax + y\|_X \\ &= \inf_{y \in M} \|a(x + y)\|_X = |a| \|q(x)\|_{X/M}. \end{aligned}$$

The triangle inequality is not much harder. Given $x, x' \in X$, and $\varepsilon > 0$, choose $y, y' \in M$ with $\|x + y\|_X \leq \|q(x)\|_{X/M} + \varepsilon$ and $\|x' + y'\|_X \leq \|q(x')\|_{X/M} + \varepsilon$. Then, since $y + y' \in M$, we get

$$\begin{aligned} \|q(x) + q(x')\|_{X/M} &= \|q(x + x')\|_{X/M} \\ &\leq \|(x + x') + (y + y')\|_X = \|(x + y) + (x' + y')\|_X \\ &\leq \|q(x)\|_{X/M} + \|q(x')\|_{X/M} + 2\varepsilon. \end{aligned}$$

What we've actually done here is to take the *quotient topology* on X/M induced by q; that is, the smallest topology on X/M making q continuous. Indeed, check that $q(B_X^o) = B_{X/M}^o$, where B_X^o denotes the *open* unit ball in X. Thus, a set is open in X/M if and only if it is the image under q of an open set in X. In particular, note that q is an *open map*.

We should also address the question of when X/M is complete. If X is complete, and if M is closed (hence complete), then it's not terribly hard to check that X/M is also complete. Indeed, suppose that (x_n) is a sequence in X for which $\sum_{n=1}^{\infty} \|q(x_n)\|_{X/M} < \infty$. For each n, choose $y_n \in M$ such that $\|x_n + y_n\|_X \leq \|q(x_n)\|_{X/M} + 2^{-n}$. Then, $\sum_{n=1}^{\infty} \|x_n + y_n\|_X < \infty$ and hence, $\sum_{n=1}^{\infty} (x_n + y_n)$ converges in X. Thus, by continuity, $\sum_{n=1}^{\infty} q(x_n + y_n) = \sum_{n=1}^{\infty} q(x_n)$ converges in X/M.

Since we're using the quotient topology, it's easy to check that a linear map $S : X/M \to Y$ is continuous if and only if Sq is continuous. Said in other words, the continuous linear maps on X/M come from continuous linear maps on X that "factor through" X/M. That is, if $T : X \to Y$ satisfies $\ker T \supset M$, then there exists a (unique) linear map $S : X/M \to Y$ satisfying $Sq = T$ and $\|S\| = \|T\|$. In the special case where T maps a Banach space X onto a Banach space Y and $M = \ker T$, it follows that the map S is one-to-one; thus, $X/(\ker T)$ is isomorphic to Y.

Finally, it's a natural question to ask whether the quotient space X/M is actually isomorphic to a subspace of X. Now, in the vector space setting, it's easy to see that if N is an algebraic complement of M in X, then X/M can be identified with N. Thus, it should come as no surprise that X/M is isomorphic to a subspace of X whenever M is complemented in X.

If X is a Banach space with $X = M \oplus N$ and if we write $Q : X \to X$ for the projection with kernel M, then our earlier observations show that $X/(\ker Q) = X/M$ is isomorphic to N. Conversely, if the quotient map $q : X \to X/M$ is an isomorphism on some closed subspace N of X, then $Q = (q|_N)^{-1}q$, considered as a map from X to X, defines a projection with range N.

In the special case of linear functionals, these observations tell us that $(X/M)^* = M^\perp$ (the annihilator of M in X^*). That is, the dual of X/M can be identified with the functionals in X^* that vanish on M. Indeed, if $f : X/M \to \mathbb{R}$ is a continuous linear functional, then $g = fq : X \to \mathbb{R}$ is a continuous linear functional that vanishes on M and satisfies $\|g\| = \|f\|$. Conversely, if $g : X \to \mathbb{R}$ is continuous and linear and satisfies $\ker g \supset M$, then, in fact, $\ker g = M$ (since they're both codimension one) and, hence, there exists a (unique) continuous, linear functional $f : X/M \to \mathbb{R}$ satisfying $g = fq$ and $\|g\| = \|f\|$. In either case, it's not hard to see that the correspondence $f \leftrightarrow g$ is linear and, hence, an isometry between $(X/M)^*$ and $M^\perp$. Alternatively, check that the adjoint of the quotient map $q : X \to X/M$ is an isometry from $(X/M)^*$ into X^* with range $M^\perp$.

A similar observation is that $M^* = X^*/M^\perp$. In other words, each continuous linear functional on M can be considered as a functional on X (thanks to Hahn–Banach); those functionals in X^* that agree on M, that is, functionals differing by an element of $M^\perp$, are simply identified for the purposes of computing M^*. Alternatively, the adjoint of the inclusion $i : M \to X$ is a (quotient) map from X^* onto M^* with kernel $M^\perp$ (in fact, i^* is the restriction map, $i^*(f) = f|_M$). Thus, $M^* = X^*/M^\perp$.

Lastly, if X is a Banach space with $X = M \oplus N$, then X^* is isomorphic to $M^* \oplus N^*$. Specifically, if $P : X \to X$ is the projection with range M and kernel N, it's not hard to see that $P^* : X^* \to X^*$ is again a projection with range $N^\perp$ and kernel $M^\perp$; that is, $X^* = N^\perp \oplus M^\perp$. Thus, $M^* = X^*/M^\perp$ is isomorphic to $N^\perp$ and, likewise, N^* is isomorphic to $M^\perp$.

A Curious Application

We round out our discussion of preliminary topics by presenting a curious result, due to Dixmier [37], that highlights many of the ideas from this chapter.

Theorem 2.2. *If* $i : X \to X^{**}$ *and* $j : X^* \to X^{***}$ *denote the canonical embeddings, then* $ji^* : X^{***} \to X^{***}$ *is a projection with range isometric to* X^* *and kernel isometric to* $i(X)^{\perp}$, *the annihilator of* $i(X)$ *in* X^{***}. *In short,* X^* *is always complemented in* X^{***}.

Proof. Note that $i^* : X^{***} \to X^*$ and, hence, $i^*j : X^* \to X^*$. To show that ji^* is a projection, it suffices to show that i^*j is the identity on X^*, for then we would have $(ji^*)(ji^*) = j(i^*j)i^* = ji^*$; that is, $(ji^*)^2 = ji^*$. So, given $x^* \in X^*$, let's compute the action of $i^*j(x^*)$ on a typical $x \in X$:

$$
\begin{aligned}
(i^*j(x^*))(x) &= \langle x, i^*j(x^*) \rangle \\
&= \langle i(x), j(x^*) \rangle \\
&= \langle x^*, i(x) \rangle \\
&= \langle x, x^* \rangle \\
&= x^*(x).
\end{aligned}
$$

Thus $i^*j(x^*) = x^*$ and, hence, i^*j is the identity on X^*.

It's not hard to see that i^* is onto; thus, the range of ji^* is $j(X^*)$, which is plainly isometric to X^*. Along similar lines, since j is one-to-one, it's easy to see that $\ker(ji^*) = \ker i^*$. Finally, general principles tell us that $\ker i^* = (\text{range } i)^{\perp} = i(X)^{\perp}$, the annihilator of $i(X)$ in X^{***}. $\square$

Notes and Remarks

There are many excellent books on functional analysis that offer more complete details for the topics in this chapter. See, for example, Bollobás [18], Conway [26], DeVito [30], Dunford and Schwartz [42], Holmes [70], Jameson [76], Järvinen [78], Megginson [100], Royden [128], Rudin [129], or Yosida [148]. Megginson's book, in particular, has a great deal on adjoints, projections, and quotients.

Exercises

Given normed spaces X and Y, we write $B(X, Y)$ to denote the space of bounded linear operators $T : X \to Y$ endowed with the operator norm (2.1).

1. Given a nonzero vector x in a normed space X, show that there exists a norm one functional $f \in X^*$ satisfying $|f(x)| = \|x\|$. On the other hand, give an example of a normed space X and a norm one linear functional $f \in X^*$ such that $|f(x)| < \|x\|$ for every $0 \neq x \in X$.

2. Let Y be a subspace of a normed linear space X and let $T \in B(Y, \mathbb{R}^n)$. Prove that T extends to a map $\tilde{T} \in B(X, \mathbb{R}^n)$ with $\|\tilde{T}\| = \|T\|$. [Hint: Hahn–Banach.]

3. Prove that every proper subspace M of a normed space X has empty interior. If M is a finite dimensional subspace of an infinite dimensional normed space X, conclude that M is nowhere dense in X.

4. Prove that $B(X, Y)$ is complete whenever Y is complete.

5. If Y is a dense linear subspace of a normed space X, show that $Y^* = X^*$, isometrically.

6. Prove Riesz's lemma: Given a closed subspace Y of a normed space X and an $\varepsilon > 0$, there is a norm one vector $x \in X$ such that $\|x - y\| > 1 - \varepsilon$ for all $y \in Y$. If X is infinite dimensional, use Riesz's lemma to construct a sequence of norm one vectors (x_n) in X satisfying $\|x_n - x_m\| \geq 1/2$ for all $n \neq m$.

7. Given linear functionals f and $(g_i)_{i=1}^n$ on a vector space, prove that $\ker f \supset \bigcap_{i=1}^n \ker g_i$ if and only if $f = \sum_{i=1}^n a_i g_i$ for some $a_1, \ldots, a_n \in \mathbb{R}$.

8. Given $T \in B(X, Y)$, show that $\ker T = {}^\perp(\text{range } T^*)$, the annihilator of range T^* in X, and that $\ker T^* = (\text{range } T)^\perp$, the annihilator of range T in Y^*.

9. Given $T \in B(X, Y)$, show that T^{**} is an extension of T to X^{**} in the following sense: If $i : X \to X^{**}$ and $j : Y \to Y^{**}$ denote the canonical embeddings, prove that $T^{**}(i(x)) = j(T(x))$. In short, $T^{**}(\hat{x}) = \widehat{Tx}$.

10. Let S be a dense linear subspace of a Banach space X, and let $T : S \to Y$ be a continuous linear map, where Y is also a Banach space. Show that T extends uniquely to a continuous linear map $\tilde{T} : X \to Y$, defined on all of X, and that $\|\tilde{T}\| = \|T\|$. Moreover, if T is an isometry, show that $\tilde{T}$ is again an isometry.

11. Let $(X\| \cdot \|)$ be a Banach space, and suppose that $\|\| \cdot \|\|$ is another norm on X satisfying $\|\|x\|\| \leq \|x\|$ for every $x \in X$. If $(X, \|\| \cdot \|\|)$ is complete, prove that there is a constant $c > 0$ such that $c\|x\| \leq \|\|x\|\|$ for every $x \in X$.

12. Let $M = \{(x, 0) : x \in \mathbb{R}\} \subset \mathbb{R}^2$. Show that there are uncountably many subspaces N of $\mathbb{R}^2$ such that $\mathbb{R}^2 = M \oplus N$.

13. Let M be a finite-dimensional subspace of a normed linear space X. Show that there is a closed subspace N of X with $X = M \oplus N$. In fact, if M is nontrivial, then there are infinitely many distinct choices for N. [Hint: Given a basis $x_1, \ldots, x_n$ for M, find $f_1, \ldots, f_n \in X^*$ with $f_i(x_j) = \delta_{i,j}$.]

14. Let M and N be closed subspaces of a Banach space X with $M \cap N = \{0\}$. Prove that $M + N$ is closed in X if and only if there is a constant $C < \infty$ such that $\|x\| \le C \|x + y\|$ for every $x \in M$, $y \in N$.

15. Let M be a closed finite-codimensional subspace of a normed space X. Show that there is a closed subspace N of X with $X = M \oplus N$.

16. Let M and N be closed subspaces of a normed space X, each having the same finite codimension. Show that M and N are isomorphic.

17. Let $P : X \to X$ be a continuous linear projection with range Y, and let $Q : X \to X$ be continuous and linear. If Q satisfies $PQ = Q$ and $QP = P$, show that Q is also a projection with range Y.

18. A bounded linear map $U : X \to X$ is called an *involution* if $U^2 = I$. If U is an involution, show that $P = \frac{1}{2}(U + I)$ is a projection. Conversely, if $P : X \to X$ is a bounded projection, then $U = 2P - I$ is an involution. In either case, P and U fix the same closed subspace $Y = \{x : Px = x\} = \{x : Ux = x\}$.

19. A projection P on a Hilbert space H is said to be an *orthogonal projection* if range P is the orthogonal complement of ker P; that is, if and only if $H = (\ker P) \oplus (\text{range } P)$. Prove that P is an orthogonal projection if and only if P is *self-adjoint*; that is, if and only if $P^* = P$.

20. Let M be a closed subspace of a normed space X. If both M and X/M are Banach spaces, then so is X. We might say that completeness is a "three space property" (if it holds in any two of the spaces X, M, or X/M, then it also holds in the third). Is separability, for example, a three space property?

21. Let $T : X \to Y$ be a bounded linear map from a normed space X onto a normed space Y. If M is a closed subspace of ker T, then there is a (unique) bounded linear map $\widetilde{T} : X/M \to Y$ such that $T = \widetilde{T}q$, where q is the quotient map. Moreover, $\|\widetilde{T}\| = \|T\|$.

22. If X and Y are Banach spaces, and if $T : X \to Y$ is a bounded linear map onto all of Y, then $X/\ker T$ is isomorphic to Y.

23. Let X and Y be Banach spaces, and let $T \in B(X, Y)$. Then, the following are equivalent:

(i) $X/\ker T$ is isomorphic to range T.

(ii) Range T is closed in Y.

(iii) There is a constant $C < \infty$ such that $\inf\{\|x - y\| : y \in \ker T\} \le C\|Tx\|$ for all $x \in X$.

24. Let M be a closed subspace of a Banach space X and let $q : X \to X/M$ be the quotient map. Prove that $q(B_X^o) = B_{X/M}^o$, where B_X^o is the open unit ball in X and $B_{X/M}^o$ is the open unit ball in X/M.

25. We say that $T \in B(X, Y)$ is a *quotient map* if $T(B_X^o) = B_Y^o$, where B_X^o denotes the open unit ball in X (respectively, Y). Prove that T is a quotient map if and only if $X/\ker T$ is isometric to Y.

26. Given $T \in B(X, Y)$, prove that T is a quotient map if and only if T^* is an isometry (into).

27. Let M be a closed subspace of a Banach space X and let $q : X \to X/M$ denote the quotient map. Prove that q^* is an isometry from $(X/M)^*$ into X^* with range $M^\perp$. Thus $(X/M)^*$ can be identified with $M^\perp$.

28. Let M be a closed subspace of a Banach space X and let $i : M \to X$ denote the inclusion map. Prove that i^* is a quotient map from X^* onto M^* with kernel $M^\perp$. Conclude that M^* can be identified with $X^*/M^\perp$.

29. Let M be a closed subspace of a normed space X. For any $f \in X^*$, show that $\min\{\|f - g\| : g \in M^\perp\} = \sup\{|f(x)| : x \in M, \|x\| \le 1\}$. (Please note the use of "min" in place of "inf.")

Chapter 3
Bases in Banach Spaces

Throughout, let X be a (real) normed space, and let (x_n) be a nonzero sequence in X. We say that (x_n) is a (Schauder) *basis* for X if, for each $x \in X$, there is an *unique* sequence of scalars (a_n) such that $x = \sum_{n=1}^{\infty} a_n x_n$, where the series converges in norm to x. Obviously, a basis for X is linearly independent. Moreover, any basis has *dense linear span*. That is, the subspace

$$\text{span}\{x_i : i \in \mathbb{N}\} = \left\{ \sum_{i=1}^{n} a_i x_i : a_1, \ldots, a_n \in \mathbb{R}, n \in \mathbb{N} \right\}$$

(consisting of all *finite* linear combinations) is dense in X. In fact, it's not hard to check that the set

$$\left\{ \sum_{i=1}^{n} a_i x_i : a_1, \ldots, a_n \in \mathbb{Q}, n \in \mathbb{N} \right\}$$

is dense in X.

We say that (x_n) is a *basic sequence* if (x_n) is a basis for its closed linear span, a space we denote by $[x_n]$ or $\overline{\text{span}}(x_n)$.

Example. ℓ_p, $1 \le p < \infty$, and c_0

It's nearly immediate that the sequence $e_n = (0, \ldots, 0, 1, 0, \ldots)$, where that single nonzero entry is in the nth slot, is a basis for ℓ_p, $1 \le p < \infty$, and for c_0. Given an element $x = (x_n) = \sum_{n=1}^{\infty} x_n e_n \in \ell_p$, for example, the very fact that $\sum_{n=1}^{\infty} |x_n|^p < \infty$ tells us that

$$\left\| x - \sum_{i=1}^{n} x_i e_i \right\|_p^p = \sum_{i=n+1}^{\infty} |x_i|^p \to 0$$

as $n \to \infty$. A similar argument applies to c_0.

Essentially the same argument shows that any subsequence (e_{n_k}) of (e_n) is a basic sequence in ℓ_p or c_0.

But please note that a space with a (Schauder) basis must be *separable*. Thus, ℓ_∞ cannot have a basis. In any case, it's easy to see that the calculation above fails irreparably in ℓ_∞.

Now all of the familiar separable Banach spaces are known to have a basis. This was well known to Banach [6], which led him to ask,

Does every separable Banach space have a basis?

The answer is no and was settled by Per Enflo [46] (presently at Kent State) in 1973. As you might imagine, a problem that took 40 years to solve has a lengthy solution. A more tractable question, from our point of view, is

Does every infinite dimensional Banach space contain a basic sequence?

The answer this time is yes and is due to Mazur in 1933. We'll give a proof of this fact shortly. A related question is,

Does every separable Banach space embed isometrically into a space with a basis?

Again, the answer to this question is yes and is due to Banach and Mazur [8], who showed that every separable Banach space embeds isometrically into $C[0, 1]$. We'll display a basis for $C[0, 1]$ very shortly. The juxtaposition of these various questions prompts the following observation: Even though $C[0, 1]$ has a basis, it apparently also has a closed subspace without a basis. Needless to say, none of the spaces involved in this discussion are Hilbert spaces!

A Schauder basis is not to be confused with a *Hamel basis*. A Hamel basis (e_α) for X is a set of linearly independent vectors in X that satisfy $\text{span}(e_\alpha) = X$. That is, each $x \in X$ is uniquely representable as a *finite* linear combination of e_α. It's an easy consequence of the Baire category theorem that a Hamel basis for an infinite-dimensional Banach space must, in fact, be *uncountable*. Indeed, suppose that $(e_i)_{i=1}^\infty$ is a Hamel basis for an infinite-dimensional Banach space X. Then each of the finite-dimensional spaces $X_n = \text{span}\{e_i : 1 \le i \le n\}$ is closed and, clearly, $X = \bigcup_{n=1}^\infty X_n$. But if X is infinite dimensional, then any finite-dimensional subspace of X has empty interior; thus, each X_n is nowhere dense, contradicting Baire's result. Hence, (e_i) must actually be uncountable. Hamel bases are of little value to us. Henceforth, the word "basis" will always mean "Schauder basis."

Given a basis (x_n), we define the *coordinate functionals* $x_n^* : X \to \mathbb{R}$ by $x_n^*(x) = a_n$, where $x = \sum_{i=1}^\infty a_i x_i$. It's easy to see that each x_n^* is linear and satisfies $x_n^*(x_m) = \delta_{m,n}$. The sequence of pairs (x_n, x_n^*) is said to be *biorthogonal*. (Although in truth we often just say that (x_n^*) is biorthogonal to (x_n).)

We also define a sequence of linear maps (P_n) on X by $P_n x = \sum_{i=1}^{n} x_i^*(x) x_i$. That is, $P_n x = \sum_{i=1}^{n} a_i x_i$, where $x = \sum_{i=1}^{\infty} a_i x_i$. It follows that the P_n satisfy $P_n P_m = P_{\min\{m,n\}}$. In particular, P_n is a *projection* onto $\text{span}\{x_i : 1 \le i \le n\}$. Also, since (x_n) is a Schauder basis, we have that $P_n x \to x$ in norm as $n \to \infty$ for each $x \in X$. Obviously, the x_n^* are all continuous precisely when the P_n are all continuous. (Why?) The amazing fact that all of these operations are continuous whenever X is complete is due to Banach:

Theorem 3.1. *If (x_n) is a basis for a Banach space X, then every P_n (and hence also every x_n^*) is continuous. Moreover, $K = \sup_n \|P_n\| < \infty$.*

Proof. Banach's ingenious idea is to define a new norm on X by setting

$$\|\|x\|\| = \sup_n \|P_n x\|.$$

Since $P_n x \to x$, it's clear that $\|\|x\|\| < \infty$ for any $x \in X$. The rest of the details required to show that $\|\| \cdot \|\|$ is a norm are more or less immediate.

To show that the P_n are uniformly bounded, we want to show that $\|\|x\|\| \le K\|x\|$ for some constant K (and all $x \in X$). To this end we appeal to a corollary of the Open Mapping theorem: Notice that the formal identity $i : (X, \|\| \cdot \|\|) \to (X, \| \cdot \|)$ is continuous since $\|x\| = \lim_{n\to\infty} \|P_n x\| \le \|\|x\|\|$. What we need to show is that this map has a continuous inverse. And for this we only need to show that $(X, \|\| \cdot \|\|)$ is a Banach space. That is, we need to check that X is complete under $\|\| \cdot \|\|$.

So, let (y_k) be a $\|\| \cdot \|\|$-Cauchy sequence. Then, for each n, the sequence $(P_n y_k)_{k=1}^{\infty}$ is $\| \cdot \|$-Cauchy. In fact, since

$$\|P_n y_i - P_n y_j\| \le \|\|y_i - y_j\|\|$$

for any n, the sequences $(P_n y_k)_{k=1}^{\infty}$ are $\| \cdot \|$-Cauchy *uniformly* in n. It follows that if $z_n = \lim_{k\to\infty} P_n y_k$ in $(X, \| \cdot \|)$, then $\|P_n y_k - z_n\| \to 0$, as $k \to \infty$, *uniformly* in n. And now, from this, it follows that (z_n) is $\| \cdot \|$-Cauchy:

$$\|z_n - z_m\| \le \|z_n - P_n y_k\| + \|P_n y_k - P_m y_k\| + \|P_m y_k - z_m\|.$$

We choose k so that the first and third terms on the right-hand side are small (uniformly in n and m), and then the middle term can be made small because $P_n y_k \to y_k$ as $n \to \infty$.

Now let $z = \lim_{n\to\infty} z_n$ in $(X, \| \cdot \|)$. We next show that $z = \lim_{k\to\infty} y_k$ in $(X, \|\| \cdot \|\|)$. For this it's enough to notice that $P_n z = z_n$. The key is that P_n

is continuous on the finite-dimensional space $P_m(X)$:

$$P_n(z_m) = P_n\left(\lim_{k\to\infty} P_m y_k\right)$$
$$= \lim_{k\to\infty} P_n P_m y_k$$
$$= \lim_{k\to\infty} P_{\min\{m,n\}} y_k = z_{\min\{m,n\}}.$$

What this means is that there is a single sequence of scalars (a_i) such that $z_n = \sum_{i=1}^{n} a_i x_i$, and so $z = \sum_{i=1}^{\infty} a_i x_i$ and $P_n z = z_n$ follow. Finally,

$$\||y_k - z\|| = \sup_n \|P_n y_k - z_n\| \to 0 \text{ as } k \to \infty. \quad \square$$

Of course, if $\sup_n \|P_n x\| \le K \|x\|$ for all x, then $\sup_n \|P_n\| \le K$. Also note that $|x_n^*(x)| \|x_n\| = \|P_n x - P_{n-1} x\| \le 2K \|x\|$. Thus, $1 \le \|x_n^*\| \|x_n\| \le 2K$.

The number $K = \sup_n \|P_n\|$ is called the *basis constant* of the basis (x_n). A basis with constant 1 is sometimes called a *monotone* basis. The canonical basis for ℓ_p, for example, is a monotone basis. It follows from the proof of our first theorem that any Banach space with a basis can always be given an equivalent norm under which the basis constant becomes 1. Indeed, $\||x\|| = \sup_n \|P_n x\|$ does the trick.

We next formulate a "test" for basic sequences; this, too, is due to Banach.

Theorem 3.2. *A sequence (x_n) of nonzero vectors is a basis for the Banach space X if and only if* (i) *(x_n) has dense linear span in X, and* (ii) *there is a constant K such that*

$$\left\| \sum_{i=1}^{n} a_i x_i \right\| \le K \left\| \sum_{i=1}^{m} a_i x_i \right\|$$

for all scalars (a_i) and all $n < m$.

(Thus, (x_n) is a basic sequence if and only if (ii) *holds.)*

Proof. The forward implication is clear; note that for $n < m$ we have

$$\left\| \sum_{i=1}^{n} a_i x_i \right\| = \left\| P_n \left(\sum_{i=1}^{m} a_i x_i \right) \right\| \le \left(\sup_j \|P_j\| \right) \left\| \sum_{i=1}^{m} a_i x_i \right\|.$$

Now suppose that (i) and (ii) hold and let $S = \text{span}\{x_i : i \ge 1\}$. Condition (i) tells us that S is dense in X. From here, condition (ii) does most of the work.

First, (ii) and induction tell us that the x_i are linearly independent:

$$|a_n| \|x_n\| \le 2K \left\| \sum_{i=1}^{m} a_i x_i \right\| \qquad (n < m).$$

Thus, the maps $P_n(\sum_{i=1}^{m} a_i x_i) = \sum_{i=1}^{n} a_i x_i$, $n < m$, are well-defined linear projections on S. Moreover, condition (ii) says that each P_n has norm at most K on S. Hence, each P_n *extends uniquely* to a continuous linear map on all of X. If we conserve notation and again call the extension P_n, then P_n is still a projection and still satisfies $\|P_n\| \leq K$. We now only need to show that $P_n x \to x$ for each $x \in X$.

Given $x \in X$ and $\varepsilon > 0$, let $s = \sum_{i=1}^{m} a_i x_i \in S$ with $\|x - s\| < \varepsilon$. Then, for $n > m$,

$$\|x - P_n x\| \leq \|x - s\| + \|s - P_n s\| + \|P_n s - P_n x\|$$
$$\leq (1 + \|P_n\|) \cdot \varepsilon \leq (K + 1) \cdot \varepsilon. \quad \square$$

We conclude this first pass at bases in Banach spaces with two classical examples, both due to J. Schauder [131, 132].

Schauder's Basis for $C[0, 1]$

Schauder began the formal theory of bases in Banach spaces in 1927 by offering up a basis for $C[0, 1]$ that now bears his name. Rather than try to give an analytical definition of the sequence, consider the following pictures:

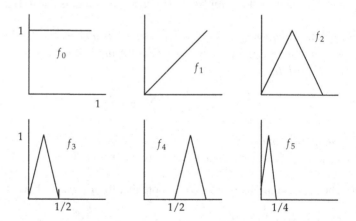

If we enumerate the dyadic rationals in the order $t_0 = 0$, $t_1 = 1$, $t_2 = 1/2$, $t_3 = 1/4$, $t_4 = 3/4$, and so on, notice that we have $f_n(t_n) = 1$ and $f_k(t_n) = 0$ for $k > n$. This easily implies that the f_n are linearly independent. Moreover, it's now easy to see that span$(f_0, \ldots, f_{2^m})$ is the set of all continuous, piecewise linear or "polygonal" functions with "nodes" at the dyadic rationals $k/2^m$, $k = 0, 1, \ldots, 2^m$. Indeed, the set of all such polygonal functions is clearly a vector space of dimension 2^{m+1} that contains the 2^{m+1} linearly independent functions f_k, $k = 0, \ldots, 2^m$. Thus, the two spaces must coincide. Since the dyadic rationals are dense in [0, 1], it's not hard to see that the f_n have dense linear span. Thus, (f_n) is a viable candidate for a basis for $C[0, 1]$.

If we set $p_n = \sum_{k=0}^{n} a_k f_k$, then

$$\|p_n\|_\infty = \max_{0 \le k \le n} |p_n(t_k)|$$

because p_n is a polygonal function with nodes at $t_0, \ldots, t_n$. And if we set $p_m = \sum_{k=0}^{m} a_k f_k$ for $m > n$, then we have $p_m(t_k) = p_n(t_k)$ for $k \le n$ because $f_j(t_k) = 0$ for $j > n \ge k$. Hence, $\|p_n\|_\infty \le \|p_m\|_\infty$. This implies that (f_n) is a normalized basis for $C[0, 1]$ with basis constant $K = 1$.

Consequently, each $f \in C[0, 1]$ can be (uniquely) written as a uniformly convergent series $f = \sum_{k=0}^{\infty} a_k f_k$. But notice, please, that $\sum_{k=n+1}^{\infty} a_k f_k$ *vanishes* at each of the nodes $t_0, \ldots, t_n$. Thus, $P_n f = \sum_{k=0}^{n} a_k f_k$ must *agree* with f at $t_0, \ldots, t_n$; that is, $P_n f$ is the interpolating polygonal approximation to f with nodes at $t_0, \ldots, t_n$. Clearly, $\|P_n f\|_\infty \le \|f\|_\infty$.

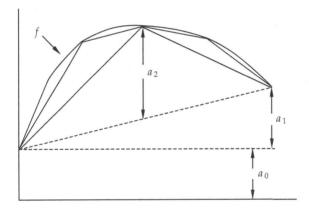

$$f = a_0 f_0 + a_1 f_1 + a_2 f_2 + \cdots$$

It's tempting to imagine that the linearly independent functions t^n, $n = 0, 1, 2, \ldots$, might form a basis for $C[0, 1]$. After all, the Weierstrass theorem tells us that the linear span of these functions is dense in $C[0, 1]$. But a moment's reflection will convince you that not every function in $C[0, 1]$ has a *uniformly* convergent power series expansion – for example, your favorite function that is not differentiable at 0. Nevertheless, as we'll see in the next chapter, $C[0, 1]$ does admit a basis consisting entirely of polynomials.

The Haar System

The Haar system $(h_n)_{n=0}^{\infty}$ on $[0, 1]$ is defined by $h_0 \equiv 1$ and, for $k = 0, 1,$ $2, \ldots$ and $i = 0, 1, \ldots, 2^k - 1$, by $h_{2^k+i}(x) = 1$ for $(2i - 2)/2^{k+1} \leq x < (2i - 1)/2^{k+1}$, $h_{2^k+i}(x) = -1$ for $(2i - 1)/2^{k+1} \leq x < 2i/2^{k+1}$, and $h_{2^k+i}(x) = 0$ otherwise. A picture might help:

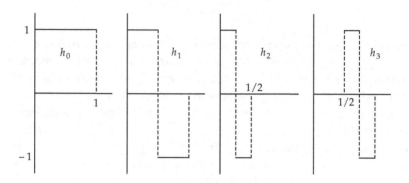

As we'll see, the Haar system is an *orthogonal* basis for $L_2[0, 1]$. Each h_n, $n \geq 1$, is mean-zero and, more generally, $h_n \cdot h_m$ is either 0 or $\pm h_m$ for any $n < m$. In particular, the h_n are linearly independent.

Note that the Schauder system is related to the Haar system by the formula $f_n(x) = 2^{n-1} \int_0^x h_{n-1}(t)\, dt$ for $n \geq 1$ (and $f_0 \equiv 1$). In fact, it was Schauder who proved that the Haar system forms a monotone basis for $L_p[0, 1]$ for any $1 \leq p < \infty$. Although we could give an elementary proof very similar to the one we used for Schauder's basis (see Exercise 3), it might be entertaining to give a slightly fancier proof. The proof we'll give borrows a small amount of terminology from probability.

For each $k = 0, 1, \ldots,$ let $A_k = \{[(i - 1)/2^{k+1}, i/2^{k+1}) : i = 1, \ldots, 2^{k+1}\}$.

Claim: The linear span of $h_0, \ldots, h_{2^{k+1}-1}$ is the set of all step functions based on the intervals in A_k. That is,

$$\text{span}\{h_0, \ldots, h_{2^{k+1}-1}\} = \text{span}\{\chi_I : I \in A_k\}.$$

Why? Well, clearly each $h_j \in \text{span}\{\chi_I : I \in A_k\}$ for $j < 2^{k+1}$, and $\text{span}\{\chi_I : I \in A_k\}$ has dimension 2^{k+1}. Thus, the two spaces must coincide. But it should be pointed out here that it's essential that we take 2^{k+1} functions at a time! The claim isn't true for an arbitrary batch of Haar functions $h_0, \ldots, h_m$.

This allows us to use the much simpler functions χ_I in place of the Haar functions in certain arguments. For example, it's now very easy to see that the h_n have dense linear span in $L_p[0, 1]$ for any $1 \le p < \infty$. Also, please note that the set $\{\chi_I : I \in A_k\}$ is again orthogonal in $L_2[0, 1]$ (although not the full set of χ_I for all dyadic intervals I).

In particular, if P_k is the orthogonal projection onto span$\{h_0, \ldots, h_{2^{k+1}-1}\}$, and if Q_k is the orthogonal projection onto span$\{\chi_I : I \in A_k\}$, then we must have $P_k f = Q_k f$. The savings here is that $Q_k f$ is very easy to compute. Indeed,

$$P_k f = Q_k f = \sum_{I \in A_k} \langle f, m(I)^{-1/2}\chi_I\rangle \cdot m(I)^{-1/2}\chi_I$$

$$= \sum_{I \in A_k} \left(\frac{1}{m(I)} \int_I f\right) \cdot \chi_I.$$

Thus, $P_k f$ is the *conditional expectation* of f given Σ_k, the σ-algebra generated by A_k. That is, $P_k f$ is the unique Σ_k-measurable function satisfying $\int_A f = \int_A P_k f$ for all $A \in \Sigma_k$.

Now let's see why P_k is a contraction on every L_p. First consider $f \in L_1[0, 1]$:

$$\int_0^1 |P_K f| = \sum_{I \in A_k} \left|\frac{1}{m(I)} \int_I f\right| \cdot m(I)$$

$$\le \sum_{I \in A_k} \int_I |f| = \int_0^1 |f|.$$

Now, for $1 < p < \infty$ and $f \in L_p[0, 1]$, we use Hölder's inequality:

$$\int_0^1 |P_k f|^p = \sum_{I \in A_k} \left|\frac{1}{m(I)} \int_I f\right|^p \cdot m(I)$$

$$\le \sum_{I \in A_k} m(I)^{1-p} \cdot m(I)^{p/q} \cdot \int_I |f|^p$$

$$= \sum_{I \in A_k} \int_I |f|^p = \int_0^1 |f|^p.$$

Thus, $\|P_k : L_p \to L_p\| \le 1$ for any $1 \le p < \infty$.

The argument in the general case follows easily from this special case. We leave the full details as an exercise, but here's an outline: Given n with $2^k < n < 2^{k+1}$, let P be the orthogonal projection onto the span of $h_0, \ldots, h_n$. Given $f \in L_p[0, 1]$, write $f = f\chi_I + f\chi_J$, where I and J are disjoint sets whose union is all of $[0, 1]$ such that $I \in \Sigma_{k-1}$ and $J \in \Sigma_k$

and such that $f\chi_I \in \text{span}\{h_0, \ldots, h_{2^k-1}\}$ and $f\chi_J \in \text{span}\{h_{2^k}, \ldots, h_n\} \subset \text{span}\{h_0, \ldots, h_{2^{k+1}-1}\}$. Then,

$$Pf = P(f\chi_I) + P(f\chi_J) = P_{k-1}(f\chi_I) \cdot \chi_I + P_k(f\chi_J) \cdot \chi_J.$$

It follows that

$$\|Pf\|_p^p = \|P_{k-1}(f\chi_I) \cdot \chi_I\|_p^p + \|P_k(f\chi_J) \cdot \chi_J\|_p^p$$
$$\leq \|f\chi_I\|_p^p + \|f\chi_J\|_p^p = \|f\|_p^p.$$

Notes and Remarks

The two main examples from this chapter are due to J. Schauder [131, 132] from 1927–28; however, our discussion of the Haar system owes much to the presentation in Lindenstrauss and Tzafriri [94, 95]. See also the 1982 *American Mathematical Monthly* article by R. C. James [75], which offers a very readable introduction to basis theory, as does Megginson [100]. For more specialized topics, see Diestel [33] or the books by Lindenstrauss and Tzafriri already cited.

There is a wealth of literature on bounded, orthogonal bases – especially bases consisting of continuous or analytic functions. See, for example, Lindenstrauss and Tzafriri [94, 95] and Wojtaszczyk [147].

If (f_n) is an orthogonal basis for $L_2[0, 1]$, then it is also a (monotone) Schauder basis for $L_2[0, 1]$. Moreover, a function biorthogonal to f_n is $g_n = f_n/\|f_n\|_2^2$, and, in this case, the canonical basis projection P_n coincides with the orthogonal projection onto $\text{span}\{f_1, \ldots, f_n\}$. However, the typical orthogonal basis for $L_2[0, 1]$ will not yield a basis for (nor even elements of) $L_p[0, 1]$ for $p \neq 2$. Nevertheless, it's known (cf. [95]) that the sequence $1, \cos t, \sin t, \cos 2t, \sin 2t, \ldots$ forms a basis for $L_p[0, 1]$, for $1 < p < \infty$ (but not for $p = 1$).

It is a rather curious fact that a *binormalized* Schauder basis for a separable Hilbert space H must be an orthonormal basis. Here is an elementary proof due to T. A. Cook [27]. Suppose that (x_n) is a Schauder basis for H with associated biorthogonal functionals (x_n^*) satisfying $\|x_n\| = \|x_n^*\| = 1$, and suppose that the inner product $\langle x_k, x_j \rangle \neq 0$ for some $k \neq j$. Then there is a unit vector e in the span of x_k and x_j that is orthogonal to x_k. Consequently, $|\langle x_j, e \rangle| > 0$. Now write $x_j = \langle x_j, e \rangle e + \langle x_k, x_j \rangle x_k$. Then, $1 = |\langle x_j, e \rangle|^2 + |\langle x_k, x_j \rangle|^2$ and, hence, $|\langle x_j, e \rangle| < 1$. Finally, notice that $1 = |x_j^*(x_j)| = |\langle x_j, e \rangle| \cdot |x_j^*(e)|$, which implies that $|x_j^*(e)| = 1/|\langle x_j, e \rangle| > 1$. This contradicts the fact that $\|x_j^*\| = 1$.

Exercises

1. Let X be a Banach space with basis (x_n). We know that the expression $|||x||| = \sup_n \|P_n x\|$ defines a norm on X that is equivalent to $\| \cdot \|$. Show that under $||| \cdot |||$, each P_n has norm one. That is, we can always renorm X so that (x_n) has basis constant $K = 1$.

2. Let (f_k) denote Schauder's basis for $C[0, 1]$ and let (t_k) denote the associated enumeration of the dyadic rationals. If $f \in C[0, 1]$ is written as $f = \sum_{k=0}^{\infty} a_k f_k$, prove that $a_n = f(t_n) - \sum_{k=0}^{n-1} a_k f_k(t_n)$.

3. Here's an outline of an elementary proof that the Haar system forms a monotone basis in every $L_p[0, 1]$, $1 \le p < \infty$.
   (a) Since $\sum_{i=0}^{n} a_i h_i$ and $\sum_{i=0}^{n+1} a_i h_i$ differ only on the support of h_{n+1}, conclude that we need to prove the inequality $|a + b|^p + |a - b|^p \ge 2|a|^p$ for all scalars a, b.
   (b) The function $f(x) = |x|^p$ satisfies $f(x) + f(y) \ge 2f((x + y)/2)$ for all x, y. [Hint: $f'' \ge 0$; hence, f' is increasing.]

4. If (x_n) is a basis for a Banach space X, under what circumstances is $(x_n/\|x_n\|)$ also a basis? In other words, can we always assume that the basis vectors are *normalized*?

5. If (x_n) is a basis for a Banach space X, under what circumstances can we renormalize so as to have $\|x_n\| = \|x_n^*\| = 1$ for all n?

6. Let (f_n) be a disjointly supported, norm one sequence in $L_p(\mu)$. Show that $\sum_{n=1}^{\infty} a_n f_n$ converges in $L_p(\mu)$ if and only if $\sum_{n=1}^{\infty} |a_n|^p < \infty$. What, if anything, is the analogue of this result when $p = \infty$? What if L_p is replaced by $C[0, 1]$ or c_0?

7. In any of the spaces ℓ_p, $1 < p < \infty$, or c_0, show that we have $e_n \overset{w}{\longrightarrow} 0$. Is the same true for $p = 1$? $p = \infty$?

8. Define $x_n = e_1 + \cdots + e_n$ in c_0. Is (x_n) a basis for c_0? What is (x_n^*) in this case? Is (x_n^*) a basis for ℓ_1?

9. Prove that a normed space X is separable if and only if there is a sequence (x_n) in X such that span(x_n) is dense in X.

10. Let X be a separable normed linear space. If E is any closed subspace of X, show that there is a sequence of norm one functionals (f_n) in X^* such that $d(x, E) = \sup_n |f_n(x)|$ for all $x \in X$. Conclude that $E = \bigcap_{n=1}^{\infty} \ker f_n$. [Hint: Given (x_n) dense in X, use the Hahn–Banach theorem to choose f_n so that $f_n = 0$ on E and $f_n(x_n) = d(x_n, E)$.]

Chapter 4
Bases in Banach Spaces II

We'll stick to the same notation throughout with just a few exceptions. Unless otherwise specified, all spaces are infinite-dimensional Banach spaces. Given a sequence (x_n) in a Banach space X, we'll use the shorthand $[x_n]$ to denote the closed linear span of (x_n). Lastly, (e_n) denotes the usual basis for ℓ_p, $1 \leq p < \infty$, or c_0.

A Wealth of Basic Sequences

We begin with a construction due to Mazur (cf., e.g. [33] or [114]) showing that every infinite-dimensional Banach space contains a basic sequence. The proof features an ingenious application of Banach's criterion for basic sequences that is of some interest in its own right.

Proposition 4.1. *Let F be a finite-dimensional subspace of an infinite-dimensional normed space X. Then, given $\varepsilon > 0$, there is an $x \in X$ with $\|x\| = 1$ such that*

$$\|y\| \leq (1 + \varepsilon)\|y + \lambda x\|$$

for all $y \in F$ and all scalars λ.

Proof. Let $0 < \varepsilon < 1$. Recall that, since F is finite dimensional, the set $S_F = \{y \in F : \|y\| = 1\}$ is *compact* in X. Thus, we can choose a finite $\varepsilon/2$-net $y_1, \ldots, y_k$ for S_F; that is, each $y \in S_F$ is within $\varepsilon/2$ of some y_i. Now, for each i, choose a norm one functional $y_i^* \in X^*$ such that $y_i^*(y_i) = 1$.

We want to find a vector x that is, in a sense, "perpendicular" to F. The next best thing, for our purposes, is to choose any norm one x with $y_1^*(x) = \cdots = y_k^*(x) = 0$. How is this possible? Well, $\bigcap_{i=1}^{k} \ker y_i^*$ is a subspace of X of *finite codimension* and so must contain a nonzero vector. The claim is that any such norm one x will do. To see this, choose $y \in S_F$, any scalar $\lambda \in \mathbb{R}$,

34

and estimate

$$\|y + \lambda x\| \geq \|y_i + \lambda x\| - \|y - y_i\|$$
$$\geq \|y_i + \lambda x\| - \varepsilon/2, \quad \text{for some } i$$
$$\geq y_i^*(y_i + \lambda x) - \varepsilon/2$$
$$= 1 - \varepsilon/2 \geq \frac{1}{1 + \varepsilon}.$$

Thus, $\|y\| \leq (1 + \varepsilon)\|y + \lambda x\|$, for all λ, whenever $\|y\| = 1$. Since the inequality is homogeneous (λ being arbitrary), this is enough. $\quad\square$

Corollary 4.2. *Every infinite-dimensional Banach space contains a closed subspace with a basis.*

Proof. Let X be an infinite-dimensional Banach space. Given $\varepsilon > 0$, choose a sequence of positive numbers (ε_n) with $\prod_{n=1}^{\infty}(1 + \varepsilon_n) \leq 1 + \varepsilon$.

We next construct a basic sequence, inductively, by repeated application of Mazur's lemma. To begin, choose any norm one $x_1 \in X$. Now, choose a norm one vector x_2 so that

$$\|y\| \leq (1 + \varepsilon_1)\|y + \lambda x_2\|$$

for all $y \in [x_1]$ and all scalars λ. Next, choose a vector x_3 of norm one so that

$$\|y\| \leq (1 + \varepsilon_2)\|y + \lambda x_3\|$$

for all $y \in [x_1, x_2]$ and all scalars λ. Choose x_4 so that.... Well, you get the picture.

The sequence (x_n) so constructed is a basic sequence with basis constant at most $K = \prod_{n=1}^{\infty}(1 + \varepsilon_n) \leq 1 + \varepsilon$. $\quad\square$

Disjointly Supported Sequences in L_p and ℓ_p

In preparation for later, let's consider an easy source for basic sequences: Disjointly supported sequences in L_p, ℓ_p, or c_0. In case it's not clear, the support of a function $f \in L_p(\mu)$ is the set $\{f \neq 0\}$. Two functions $f, g \in L_p(\mu)$ are thus disjointly supported if $\{f \neq 0\} \cap \{g \neq 0\} = \emptyset$; that is, if $f \cdot g = 0$. In ℓ_p or c_0 this reads: $x = (x_n)$ and $y = (y_n)$ are disjointly supported if $x_n y_n = 0$ for all n.

Lemma 4.3. *Let (f_n) be a sequence of disjointly supported nonzero vectors in $L_p(\mu)$, $1 \leq p < \infty$. Then (f_n) is a basic sequence in $L_p(\mu)$. Moreover, $[f_n]$ is isometric to ℓ_p and is complemented in $L_p(\mu)$ by a norm one projection.*

Proof. The linear span of (f_n) is unaffected if we replace each f_n by $f_n/\|f_n\|_p$. Thus, we may assume that each f_n is norm one.

Next, since the f_k are disjointly supported, we have

$$\left\|\sum_{k=n}^{m} a_k f_k\right\|_p^p = \int \left|\sum_{k=n}^{m} a_k f_k\right|^p d\mu = \sum_{k=n}^{m} |a_k|^p \int |f_k|^p d\mu = \sum_{k=n}^{m} |a_k|^p$$

for any scalars (a_k). This tells us that $\sum_{n=1}^{\infty} a_n f_n$ converges in $L_p(\mu)$ if and only if $\sum_{n=1}^{\infty} |a_n|^p < \infty$. (Why?) Thus, the map $e_n \mapsto f_n$ extends to a linear isometry from ℓ_p onto $[f_n]$. In particular, it follows that (f_n) is a basic sequence in $L_p(\mu)$.

Lastly, the existence of a projection is easy. A sequence (g_n), biorthogonal to (f_n), is clearly given by $g_n = (\operatorname{sgn} f_n)|f_n|^{p-1}$. The g_n are disjointly supported, norm one functions in $L_q(\mu)$, where q is the conjugate exponent to p. Moreover, g_n has the same support as f_n, namely, $A_n = \{g_n \neq 0\} = \{f_n \neq 0\}$. Now consider

$$Pf = \sum_{n=1}^{\infty} \langle f, g_n \rangle f_n = \sum_{n=1}^{\infty} \left(\int_{A_n} f g_n\right) f_n.$$

Clearly, P is the identity on $[f_n]$. To show that P is norm one, we essentially repeat our first calculation:

$$\|Pf\|_p^p = \sum_{n=1}^{\infty} \left|\int_{A_n} f g_n\right|^p \leq \sum_{n=1}^{\infty} \int_{A_n} |f|^p \leq \|f\|_p^p$$

(the inequality coming from Hölder's inequality). □

It's also true that a disjointly supported sequence in c_0 spans an isometric copy of c_0. The proof is a simple modification of the one we've just given. Rather than repeat the proof in this case, let's settle for pointing out two such modifications. First, if (y_n) is a sequence of disjointly supported norm one vectors in c_0, then $\sum_{n=1}^{\infty} a_n y_n$ converges in c_0 if and only if $a_n \to 0$ and, in this case,

$$\left\|\sum_{n=1}^{\infty} a_n y_n\right\|_{\infty} = \sup_n \|a_n y_n\|_{\infty} = \sup_n |a_n|.$$

The natural sequence of biorthogonal functionals in this case is simply an appropriately chosen subsequence of (e_k) up to sign. Specifically, given $y_n \in c_0$, choose k_n so that y_n attains its norm in its k_nth coordinate; that is, $\|y_n\|_{\infty} = |\langle y_n, e_{k_n}\rangle| = \langle y_n, \pm e_{k_n}\rangle$.

The key fact in our last lemma is that $\|\cdot\|_p^p$ is additive across sums of disjointly supported functions. But if we're willing to settle for "isomorphic" in place of "isometric," then all we need is "almost additive" in place of "additive" or "almost disjointly supported" in place of "disjointly supported." This suggests the following generalization:

Lemma 4.4. *Let (f_n) be a sequence of norm one functions in $L_p(\mu)$. Suppose that there exists a sequence of disjoint, measurable sets (A_n) such that $\int_{A_n^c} |f_n|^p \, d\mu < \varepsilon_n^p$, where $\varepsilon_n \to 0$ "fast enough." Then (f_n) is a basic sequence in $L_p(\mu)$. Moreover, $[f_n]$ is isomorphic to ℓ_p and complemented in $L_p(\mu)$.*

Proof. From our previous lemma, we know how to deal with the disjointly supported sequence $\tilde{f}_n = f_n \chi_{A_n}$. The idea here is that (f_n) is a "small perturbation" of $(\tilde{f}_n)$. By design, $\| f_n - \tilde{f}_n \|_p^p = \int_{A_n^c} |f_n|^p < \varepsilon_n^p$; thus,

$$\left\| \sum_n a_n f_n - \sum_n a_n \tilde{f}_n \right\|_p \leq \sum_n |a_n| \| f_n - \tilde{f}_n \|_p \leq \sum_n \varepsilon_n |a_n|.$$

This gives us a hint as to what "fast enough" should mean: Given $0 < \varepsilon < 1/2$, let's suppose that $\sum_{n=1}^\infty \varepsilon_n < \varepsilon$. Then,

$$\sum_n \varepsilon_n |a_n| \leq \left(\sum_n \varepsilon_n \right) \cdot \sup_n |a_n| \leq \varepsilon \cdot \left(\sum_n |a_n|^p \right)^{1/p}.$$

From this, and our previous lemma, it follows that

$$\left\| \sum_n a_n f_n \right\|_p \leq \left\| \sum_n a_n \tilde{f}_n \right\|_p + \varepsilon \cdot \left(\sum_n |a_n|^p \right)^{1/p}$$

$$\leq (1 + \varepsilon) \cdot \left(\sum_n |a_n|^p \right)^{1/p}.$$

If we can establish a similar lower estimate, we will have shown that $[f_n]$ is isomorphic to ℓ_p. But,

$$\left\| \sum_n a_n \tilde{f}_n \right\|_p^p = \sum_n |a_n|^p \int_{A_n} |f_n|^p \, d\mu \geq \sum_n (1 - \varepsilon_n^p) |a_n|^p,$$

and $1 - \varepsilon_n^p \geq (1 - \varepsilon_n)^p \geq (1 - \varepsilon)^p$; hence,

$$\left\| \sum_n a_n f_n \right\|_p \geq \left\| \sum_n a_n \tilde{f}_n \right\|_p - \varepsilon \cdot \left(\sum_n |a_n|^p \right)^{1/p}$$

$$\geq (1 - 2\varepsilon) \cdot \left(\sum_n |a_n|^p \right)^{1/p}.$$

To find a bounded projection onto $[f_n]$, we now mimic this idea to show that our "best guess" is another "small perturbation" of the projection given in the previous lemma. The map that should work is

$$Pf = \sum_{n=1}^{\infty} \left(\|\tilde{f}_n\|_p^{-p} \int (\operatorname{sgn}\tilde{f}_n)|\tilde{f}_n|^{p-1} f \right) f_n.$$

Rather than give the rest of the details now, we'll save our strength for a more general result that we'll see later. □

Equivalent Bases

Two basic sequences (x_n) and (y_n) (in possibly different spaces!) are said to be *equivalent* if $\sum_{i=1}^{\infty} a_i x_i$ and $\sum_{i=1}^{\infty} a_i y_i$ converge or diverge together. A straight-forward application of the Closed Graph theorem allows us to rephrase this condition: (x_n) and (y_n) are equivalent if there exists a constant $0 < C < \infty$ such that

$$C^{-1} \left\| \sum_{i=1}^{\infty} a_i y_i \right\|_Y \leq \left\| \sum_{i=1}^{\infty} a_i x_i \right\|_X \leq C \left\| \sum_{i=1}^{\infty} a_i y_i \right\|_Y \tag{4.1}$$

for all scalars (a_i). That is, (x_n) and (y_n) are equivalent if and only if the basis-to-basis map $x_i \mapsto y_i$ extends to an isomorphism between $[x_n]$ and $[y_n]$. Thus, we may (and, in practice, will) take (4.1) as the defining statement.

Stated in these terms, the condition takes on new significance: If we start with a basic sequence (x_n), and if we find that some sequence (y_n) satisfies (4.1), then (y_n) must also be a basic sequence – and is equivalent to (x_n), of course. Indeed, if (x_n) has basis constant K, and if (y_n) satisfies (4.1), then (y_n) has basis constant at most $C^2 K$:

$$\left\| \sum_{i=1}^{n} a_i y_i \right\| \leq C \left\| \sum_{i=1}^{n} a_i x_i \right\| \leq CK \left\| \sum_{i=1}^{m} a_i x_i \right\| \leq C^2 K \left\| \sum_{i=1}^{m} a_i y_i \right\|.$$

As a particular example, a sequence (x_n) is equivalent to the usual basis for ℓ_p if and only if

$$C^{-1} \left(\sum_{i=1}^{\infty} |a_i|^p \right)^{1/p} \leq \left\| \sum_{i=1}^{\infty} a_i x_i \right\|_X \leq C \left(\sum_{i=1}^{\infty} |a_i|^p \right)^{1/p} \tag{4.2}$$

for some constant C and all scalars (a_i).

If (4.1) holds, we sometimes say that (x_n) and (y_n) are *C-equivalent*. Thus we might say that a disjointly supported sequence of norm one vectors in

L_p is 1-equivalent to the usual basis of ℓ_p. Or, to paraphrase Lemma 4.4, an "almost disjoint" sequence in L_p is $(1 + \varepsilon)$-equivalent to the ℓ_p basis. As another example, note that any orthonormal basis in a separable Hilbert space is 1-equivalent to the usual basis of ℓ_2.

Next, let's generalize the content of Lemma 4.4.

Theorem 4.5 (The Principle of Small Perturbations). *Let (x_n) be a normalized basic sequence in a Banach space X with basis constant K, and suppose that (y_n) is a sequence in X with $\sum_{n=1}^{\infty} \|x_n - y_n\| = \delta$.*

(i) *If $2K\delta < 1$, then (y_n) is a basic sequence equivalent to (x_n).*

(ii) *If $[x_n]$ is complemented by a bounded projection $P : X \to X$, and if $8K\delta\|P\| < 1$, then $[y_n]$ is also complemented in X.*

Proof. We begin with a "microlemma." For any sequence of scalars (a_i) and any n, first recall that

$$|a_n| = \|a_n x_n\| = \|P_n x - P_{n-1} x\| \le 2K \|x\|,$$

where $x = \sum_n a_n x_n$. That is, the coordinate functionals all have norm at most $2K$. In particular, we always have

$$\frac{1}{2K} \sup_n |a_n| \le \left\| \sum_{n=1}^{\infty} a_n x_n \right\| \le \sum_{n=1}^{\infty} |a_n|.$$

Now, on with the proof.... To begin, notice that

$$\left\| \sum_n a_n x_n - \sum_n a_n y_n \right\| \le \sum_n |a_n| \|x_n - y_n\| \le \delta \sup_n |a_n|$$

$$\le 2K\delta \left\| \sum_n a_n x_n \right\|.$$

Thus,

$$(1 - 2K\delta) \left\| \sum_n a_n x_n \right\| \le \left\| \sum_n a_n y_n \right\|$$

$$\le (1 + 2K\delta) \left\| \sum_n a_n x_n \right\|, \tag{4.3}$$

and hence (y_n) is a basic sequence equivalent to (x_n).

In other words, we've shown that the map $T(\sum_n a_n x_n) = \sum_n a_n y_n$ is an isomorphism between $[x_n]$ and $[y_n]$. Note that (4.3) gives $\|T\| \le 1 + 2K\delta < 2$ and $\|T^{-1}\| \le (1 - 2K\delta)^{-1}$.

To prove (ii), we next note that any nontrivial projection P has $\|P\| \geq 1$, and hence the condition $8K\delta\|P\| < 1$ implies, at the very least, that $4K\delta < 1$. A bit of arithmetic will convince you that this gives us $\|x\| < 2\|y\|$, where $x = \sum_n a_n x_n$ and $y = Tx$ (that is, $\|T^{-1}\| < 2$). In particular, it follows from our "microlemma" that the coordinate functionals for the y_n have norm at most $4K$ (that is, $|a_n| \leq 4K\|y\|$ where $y = \sum_n a_n y_n$).

Next, we show that TP is an isomorphism on $Y = [y_n]$. Indeed, if $y = \sum_n a_n y_n$ and $x = \sum_n a_n x_n$, then

$$
\begin{aligned}
\|TPy - y\| &= \|TP(y - x)\| \\
&= \left\| TP \left(\sum_{n=1}^{\infty} a_n(y_n - x_n) \right) \right\| \\
&\leq \|T\|\|P\| \left(\sup_n |a_n| \right) \cdot \sum_n \|y_n - x_n\| \\
&\leq 8K\delta\|P\|\|y\| < \|y\|.
\end{aligned}
$$

It follows (see the next lemma) that $S = (TP)|_Y$ is an invertible map on Y. Hence, $Q = S^{-1}TP$ is a projection from X onto Y. $\quad\square$

Now the proof that we've just given supplies a hint as to how we might further improve the result. The "microlemma" tells us that we want $\|x_n^*\|\delta < 1$ for all n, where x_n^* is the nth coordinate functional (which, by Hahn–Banach, we can take to be an element of X^*). Or, better still, we might ask for $\sum_{n=1}^{\infty} \|x_n^*\|\|x_n - y_n\| < 1$. This sum estimates the norm of the map $S : X \to X$ defined by

$$
Sx = \sum_{n=1}^{\infty} x_n^*(x)(x_n - y_n).
$$

What is the map S doing here? Well, if we're given $x = \sum_{n=1}^{\infty} x_n^*(x)x_n$ in $[x_n]$, then the basis-to-basis map should send x into

$$
Tx = \sum_{n=1}^{\infty} x_n^*(x)y_n.
$$

Thus, at least on $[x_n]$, we have $S = I - T$. If S is "small enough," that is, if T is "close enough" to I, then T should be an isomorphism on $[x_n]$. That this is true is a useful fact in its own right and is well worth including.

Lemma 4.6. *If a linear map $S : X \to X$ on a Banach space X has $\|S\| < 1$, then $I - S$ has a bounded inverse and $\|(I - S)^{-1}\| \leq (1 - \|S\|)^{-1}$.*

Proof. The geometric series $I + S + S^2 + S^3 + \cdots$ converges in operator norm to a bounded operator U with $\|U\| \leq (1 - \|S\|)^{-1}$. That $U = (I - S)^{-1}$ follows by simply checking that $(I - S)Ux = x = U(I - S)x$ for any $x \in X$. $\square$

Given this setup, see if you can supply a proof for the following new and improved version of Theorem 4.5.

Theorem 4.7 (The Principle of Small Perturbations). *Let (x_n) be a basic sequence in a Banach space X with corresponding coordinate functionals (x_n^*). Suppose that (y_n) is a sequence in X with $\sum_{n=1}^{\infty} \|x_n^*\| \|x_n - y_n\| = \delta$.*

(i) *If $\delta < 1$, then (y_n) is a basic sequence equivalent to (x_n).*
(ii) *If $[x_n]$ is the range of a projection $P : X \to X$, and if $\delta\|P\| < 1$, then $[y_n]$ is complemented in X.*

Hint: For (ii), show that the map $A : X \to X$ defined by

$$Ax = x - Px + \sum_{n=1}^{\infty} x_n^*(Px)y_n$$

satisfies $\|I - A\| < 1$ and $Ax_n = y_n$. The projection onto $[y_n]$ is then given by $Q = APA^{-1}$.

Notes and Remarks

Mazur's construction (Proposition 4.1 and Corollary 4.2) is part of the folklore of Banach space theory and, as far as I know, was never actually published by Mazur. Instead, we know about it through "word of mouth" supplied by the pioneering early papers of Pełczyński. Pełczyński [114] claims that Proposition 4.1 was first proved by Bessaga in his thesis but credits Mazur for the idea behind both Bessaga's proof and Pełczyński's (presented here). In a later paper, Pełczyński [115] refers also to a 1959 paper by Bessaga and Pełczyński [16].

The material on disjointly supported sequences in L_p and ℓ_p is "old as the hills" and was well known to Banach. The variations offered by the notion of "almost disjointness" and the principle of small perturbations, however, are somewhat more modern and can be traced to the early work of Bessaga and Pełczyński [15]. As a simple application of the principle of small perturbations, it follows that $C[0, 1]$ has a basis consisting entirely of polynomials. The same is true of $L_p[0, 1]$.

As a final comment regarding equivalent bases, we should point out that there is no such thing as "the" basis for a given space. Granted, there are some spaces in which a "natural" basis suggests itself, ℓ_p for example, but, in general, a basis, even if one exists, is far from unique. That this is so is shown rather dramatically by the following theorem due to Pełczyński and Singer [116].

Theorem 4.8. *If X is an infinite-dimensional Banach space with a Schauder basis, then there are uncountably many mutually nonequivalent, normalized bases in X.*

Exercises

Recall that a sequence (x_n) in a normed space X converges *weakly* to $x \in X$, written $x_n \xrightarrow{w} x$, if $f(x_n) \to f(x)$ for every $f \in X^*$. It's easy to see that $x_n \xrightarrow{w} x$ if and only if $x_n - x \xrightarrow{w} 0$. A sequence tending weakly to 0 is said to be *weakly null*.

1. Let (f_n) be a sequence of disjointly supported functions in L_p, $1 < p < \infty$. Prove that $\sum_n f_n$ converges in L_p if and only if $\sum_n \|f_n\|_p^p < \infty$.

2. Let (x_n) be a disjointly supported norm one sequence in c_0. Prove that $[x_n]$ is isometric to c_0 and complemented in c_0 by a norm one projection.

3. Let (f_n) be a disjointly supported norm one sequence in $C[0, 1]$. Prove that $[f_n]$ is isometric to c_0. Is $[f_n]$ complemented in $C[0, 1]$? Will these arguments carry over to disjointly supported sequences in $L_\infty[0, 1]$?

4. Let (x_n) be a basis for a Banach space X, and let (y_n) be a sequence in a Banach space Y. Suppose that $\sum_n a_n y_n$ converges in Y whenever $\sum_n a_n x_n$ converges in X, where (a_n) is a sequence of scalars. Use the Closed Graph theorem to prove that formula (4.1) holds for some constant $0 < C < \infty$.

5. Prove Theorem 4.7.

6. Let $T : X \to X$ be a continuous linear map on a Banach space X. If T is invertible and if $S : X \to X$ is a linear map satisfying $\|T - S\| < \|T^{-1}\|^{-1}$, prove that S is also invertible. Thus, the set of invertible maps on X is open in $B(X)$.

7. In each of the spaces ℓ_p, $1 \le p < \infty$, or c_0, the standard basis (e_n) is weakly null but not norm null. In fact, the set $\{e_n : n \ge 1\}$ is norm closed. Similarly, in any Hilbert space, an orthonormal sequence (x_n) is always weakly null, whereas the set $\{x_n : n \ge 1\}$ is always norm closed.

8. Let (f_n) be a disjointly supported sequence of norm one vectors in $L_p(\mu)$, $1 < p < \infty$. Prove that $f_n \xrightarrow{w} 0$. Is the same true for $p = 1$? $p = \infty$?

9. If $T : X \to Y$ is a bounded linear map, and if $x_n \xrightarrow{w} 0$ in X, prove that $Tx_n \xrightarrow{w} 0$ in Y.

10. Suppose that X and Y are isomorphic Banach spaces and that Z is a complemented subspace of Y. Prove that X contains a complemented subspace W that is isomorphic to Z.

Chapter 5

Bases in Banach Spaces III

Recall Banach's *basis problem*: Does every separable Banach space have a basis? Although the question was ultimately answered in the negative, several positive results were uncovered along the way. For example, we might ask instead, Does every separable Banach space *embed* in a space with a basis? Or, does every Banach space contain a *subspace* with a basis; that is, does every Banach space contain a basic sequence? The answers to both of these amended problems are yes, and both were known to Banach. The first follows from the amazing fact, due to Banach and Mazur [8], that every separable Banach space embeds *isometrically* into $C[0, 1]$. One of our goals in this short course is to give a proof of this universal property of $C[0, 1]$.

The second question, the existence of basic sequences, was settled by Mazur, as we saw in the last chapter (Corollary 4.2). In this chapter we give a second solution, due to Bessaga and Pełczyński [15], which features a useful selection principle.

Our goal here is to mimic the simple case of disjointly supported sequences in ℓ_p. The only real difference is the interpretation of the word "disjoint." In a space with a basis (x_n), we could interpret "disjointly supported" to mean "having nonzero coefficients, relative to the basis (x_n), occurring over disjoint subsets of $\mathbb{N}$." That is, we could say that $x = \sum_{n=1}^{\infty} a_n x_n$ and $y = \sum_{n=1}^{\infty} b_n x_n$ are disjointly supported relative to the basis (x_n), if $a_n b_n = 0$ for all n.

Block Basic Sequences

Let (x_n) be a basic sequence in a Banach space X. Given increasing sequences of positive integers $p_1 < q_1 < p_2 < q_2 < \cdots$, let $y_k = \sum_{i=p_k}^{q_k} b_i x_i$ be any nonzero vector in the span of $x_{p_k}, \ldots, x_{q_k}$. We say that (y_k) is a *block basic sequence* with respect to (x_n). It's easy to see that (y_k) is, indeed, a basic sequence with the same basis constant as (x_n):

$$\left\| \sum_{k=1}^{n} a_k y_k \right\| = \left\| \sum_{k=1}^{n} \sum_{i=p_k}^{q_k} a_k b_i x_i \right\| \leq K \left\| \sum_{k=1}^{m} \sum_{i=p_k}^{q_k} a_k b_i x_i \right\| = K \left\| \sum_{k=1}^{m} a_k y_k \right\|.$$

If (x_n) is fixed, we'll simply call (y_k) a block basic sequence or even just a *block basis*. By way of a simple example, note that any subsequence (x_{n_k}) of a basis (x_n) is a block basis.

We next show how to "extract" basic subsequences. The method we'll use is a standard bit of trickery known as a "gliding hump" argument. You may find it helpful to draw some pictures to go along with the proof.

Lemma 5.1. *Let (x_n) be a basis for a Banach space X, and let (x_n^*) be the associated coefficient functionals. Suppose that (z_n) is a nonzero sequence in X such that $\lim_{n \to \infty} x_i^*(z_n) = 0$ for each i. Then, given $\varepsilon_k > 0$, there is a subsequence (z_{n_k}) of (z_n) and a block basic sequence (y_k), relative to (x_n), such that $\|z_{n_k} - y_k\| \leq \varepsilon_k$ for every k.*

Proof. Let $n_1 = 1$. Since $z_{n_1} = \sum_{i=1}^{\infty} x_i^*(z_{n_1}) x_i$ converges in X, we can find some $q_1 > 1$ such that

$$\left\| \sum_{i=q_1+1}^{\infty} x_i^*(z_{n_1}) x_i \right\| \leq \varepsilon_1.$$

Setting $p_1 = 1$ and $y_1 = \sum_{i=p_1}^{q_1} x_i^*(z_{n_1}) x_i$ yields $\|z_{n_1} - y_1\| \leq \varepsilon_1$.

The idea now is to find a vector z_{n_2} which is "almost disjoint" from z_{n_1}. For this we use that fact that $\lim_{n \to \infty} x_i^*(z_n) = 0$ for each i. In particular, by applying this fact to only finitely many i we can find an $n_2 > n_1$ such that

$$\left\| \sum_{i=1}^{q_1} x_i^*(z_{n_2}) x_i \right\| \leq \varepsilon_2/2.$$

(The span of finitely many x_i is just $\mathbb{R}^n$ in disguise!) Let $p_2 = q_1 + 1$ and choose $q_2 > p_2$ such that

$$\left\| \sum_{i=q_2+1}^{\infty} x_i^*(z_{n_1}) x_i \right\| \leq \varepsilon_2/2.$$

Setting $y_2 = \sum_{i=p_2}^{q_2} x_i^*(z_{n_1}) x_i$ then yields $\|z_{n_2} - y_2\| \leq \varepsilon_2$.

We continue, finding z_{n_3} "almost disjoint" from z_{n_1} and z_{n_2}, and so on. $\square$

The last proof contains a minor mistake! Did you spot it? The fly in the ointment is that we have no way of knowing whether the y_k are nonzero! This is easy to fix, though: We should insist that the z_{n_k} are bounded away from zero. The principle of small perturbations tells us what to do next:

Lemma 5.2. *Using the same notation as in Lemma 5.1, suppose that, in addition,* $\liminf_{n\to\infty} \|z_n\| > 0$. *Then,* (z_n) *has a subsequence that is basic and that is equivalent to some block basic sequence of* (x_n).

Proof. By passing to a subsequence if necessary, we may suppose that $\|z_n\| \geq \varepsilon > 0$ for all n. Now, by taking $\varepsilon_k \to 0$ "fast enough" in Lemma 5.1, the principle of small perturbations (Theorem 4.5) will apply. $\square$

By modifying our gliding hump argument just slightly, we arrive at

Corollary 5.3. *Let X be a Banach space with a basis (x_n), and let E be an infinite dimensional subspace of X. Then, E contains a basic sequence equivalent to some block basis of (x_n).*

Proof. It suffices to show that E contains a sequence of norm one vectors (z_n) such that $\lim_{n\to\infty} x_i^*(z_n) = 0$ for each i. But, in fact, we'll prove something more:

Claim: For each $m = 1, 2, \ldots$, there exists a norm one vector $z_m \in E$ such that $x_i^*(z_m) = 0$ for all $i = 1, \ldots, m$; that is, E contains a norm one vector of the form $z_m = \sum_{n=m+1}^{\infty} a_n x_n$.

How can this be? Well $\bigcap_{i=1}^{m} \ker x_i^*$ is a subspace of X of codimension at most m and so must intersect every infinite-dimensional subspace of X nontrivially. Alternatively, consider the linear map $z \mapsto (x_1^*(z), \ldots, x_m^*(z))$ from E into $\mathbb{R}^m$. Since E is infinite dimensional, this map must have a nontrivial kernel; that is, there must be some norm one $z \in E$ that is mapped to $(0, \ldots, 0)$. $\square$

Once we have shown that every separable Banach space embeds isometrically into $C[0, 1]$, an application of Corollary 5.3 will yield that every Banach space contains a basic sequence. Indeed, it would then follow that every separable Banach space contains a basic sequence (equivalent to a block basis of the Schauder basis for $C[0, 1]$). Since every Banach space obviously contains a closed separable subspace, this does the trick.

A similar approach can be used to show what we might call *the Bessaga–Pełczyński selection principle*:

Corollary 5.4. *Let* X *be a Banach space, and suppose that* (z_n) *is a sequence in* X *such that* $z_n \xrightarrow{w} 0$ *but* $\|z_n\| \nrightarrow 0$. *Then,* (z_n) *has a basic subsequence.*

Subspaces of ℓ_p and c_0

Temporarily, X will denote one of the spaces ℓ_p, $1 \le p < \infty$, or c_0. We'll use (e_n) to denote the standard basis in X and (e_n^*) to denote the associated sequence of coordinate functionals in X^*. (Note that (e_n^*) is really just (e_n) again but considered as a sequence in the dual space.) Let's summarize what we know about the (closed, infinite-dimensional) subspaces of X.

Proposition 5.5. *Let* (y_n) *be any disjointly supported, nonzero sequence in* X. *Then,* $[y_n]$ *is isometric to* X *and is complemented in* X *by a projection of norm one.*

This is immediate from Lemma 4.3.

Corollary 5.6. *Any seminormalized block basis* (y_n) *of* (e_n) *is equivalent to* (e_n). *Moreover,* $[y_n]$ *is isometric to* X *and is complemented in* X *by a projection of norm one.*

A *seminormalized* sequence (y_n) satisfies $\inf_n \|y_n\| > 0$ and $\sup_n \|y_n\| < \infty$. This assumption is needed to check the equivalence with (e_n). Finally,

Corollary 5.7. *Every infinite-dimensional subspace of* X *contains a further subspace that is isomorphic to* X *and complemented in* X.

As a brief-reminder, the proof of this last fact consists of first showing that every infinite-dimensional subspace contains an "almost disjoint" sequence of norm one vectors. Such a sequence is a small perturbation of a block basic sequence and so is "almost isometric" to X and is the range of a projection of norm "almost one." In fact, given $\varepsilon > 0$, we can find a subspace that is $(1 + \varepsilon)$-isomorphic to X and $(1 + \varepsilon)$-complemented in X.

The fact that every subspace of X contains another copy of X is just what we need to prove that each member of the family ℓ_p, $1 \le p < \infty$, and c_0 is isomorphically distinct. In fact, no space from this class is isomorphic to a subspace of another member of the class. To see this, let's first consider a special case:

Theorem 5.8. *Let* $1 \leq p < r < \infty$, *and let* $T : \ell_r \to \ell_p$ *be a bounded linear map. Then* $\|Te_n\|_p \to 0$. *In particular,* T *is not an isomorphism. The same is true of any map* $T : c_0 \to \ell_p$.

Proof. First note that $Te_n \xrightarrow{w} 0$. That is, given any $f \in \ell_q = (\ell_p)^*$, where $1/p + 1/q = 1$, the claim here is that $f(Te_n) \to 0$ as $n \to \infty$. But $f \circ T$ is an element of $\ell_s = (\ell_r)^*$, where $1/r + 1/s = 1$, and $f(Te_n) = (f \circ T)(e_n) = e_n^*(f \circ T)$ is just the nth coordinate of $f \circ T$. Thus, since $1 < s < \infty$, we must have $f(Te_n) \to 0$ as $n \to \infty$.

Here's the same proof in different words:

$$\langle Te_n, f \rangle = \langle e_n, T^*f \rangle \to 0 \quad \text{as} \quad n \to \infty$$

because $e_n \xrightarrow{w} 0$ in ℓ_r.

Now, suppose that $\|Te_n\|_p \nrightarrow 0$; that is, suppose $\liminf_{n \to \infty} \|Te_n\|_p > 0$. Then, by Lemma 5.2, some subsequence of (Te_n) is basic and is equivalent to a block basis of (e_n) in ℓ_p which, by Corollary 5.6, is in turn equivalent to (e_n). In particular, after passing to a subsequence, we can find a constant $C < \infty$ such that

$$\left\| \sum_{k=1}^{\infty} a_k e_k \right\|_p \leq C \left\| \sum_{k=1}^{\infty} a_k Te_{n_k} \right\|_p .$$

Thus,

$$\left(\sum_{k=1}^{\infty} |a_k|^p \right)^{1/p} \leq C\|T\| \left\| \sum_{n=1}^{\infty} a_k e_{n_k} \right\|_r = C\|T\| \left(\sum_{k=1}^{\infty} |a_k|^r \right)^{1/r} .$$

We've arrived at a contradiction: If this inequality were to hold for all scalars, then, in particular, we'd have $n^{1/p} \leq C\|T\| n^{1/r}$ for all n. Since $p < r$, this is impossible. Consequently, $\|Te_n\|_p \to 0$.

The proof in case $T : c_0 \to \ell_p$ is virtually identical. $\square$

With just a bit more work, we could improve this result to read: A bounded linear map $T : \ell_r \to \ell_p$, $1 \leq p < r < \infty$, or $T : c_0 \to \ell_p$ is *compact*. That is, T maps bounded sets into compact sets.

This proof of Theorem 5.8 actually shows something more: *T fails to be an isomorphism on any infinite-dimensional subspace of* ℓ_r. Indeed, each infinite-dimensional subspace of ℓ_r contains an equivalent "copy" of (e_n), and we could repeat the proof for this "copy." Or, in still other words, each infinite-dimensional subspace of ℓ_r contains an isomorphic copy of ℓ_r and the restriction of T to this copy of ℓ_r cannot be an isomorphism. A bounded linear map $T : X \to Y$ that fails to be an isomorphism on any subspace of X is said to be *strictly singular*. (Any compact operator, for example, is strictly singular, but not conversely.) Thus, every map $T : \ell_r \to \ell_p$, where $1 \leq p < r < \infty$, is strictly singular. Likewise for maps $T : c_0 \to \ell_p$. But even more is true:

Corollary 5.9. *Let X and Y be two distinct members of the family of spaces c_0 and ℓ_p, $1 \leq p < \infty$. Then, every bounded linear map $T : X \to Y$ is strictly singular. In particular, X and Y are not isomorphic.*

Proof. We consider the case $T : \ell_r \to \ell_p$ where $1 \leq r < p < \infty$ (and leave the remaining cases as an exercise).

Suppose that T is an isomorphism from a subspace of ℓ_r onto a subspace W of ℓ_p. Then there is a further subspace Z of W and an isomorphism $S : \ell_p \to Z$. But then $T^{-1}S$ is an isomorphism from ℓ_p into ℓ_r, which is impossible. $\square$

Complemented Subspaces of ℓ_p and c_0

In this section, we present Pełczyński's characterization of the complemented subspaces of ℓ_p and c_0 [113]. His proof is based on an elegant and mysterious *decomposition method*. Before we can describe the method, we'll need a few preliminary facts.

Given two Banach spaces X and Y, we can envision their sum $X \oplus Y$ as the space of all pairs (x, y), where $x \in X$ and $y \in Y$. Up to isomorphism, it doesn't much matter what norm we take on $X \oplus Y$. For example, if we write $(X \oplus Y)_p$ to denote $X \oplus Y$ under the norm $\|(x, y)\| = (\|x\|_X^p + \|y\|_Y^p)^{1/p}$, then

$$(X \oplus Y)_1 \approx (X \oplus Y)_p \approx (X \oplus Y)_\infty,$$

where "$\approx$" means "is isomorphic to." This is a simple consequence of the fact that all norms on $\mathbb{R}^2$ are equivalent. As a particular example, note that under any norm

$$\ell_p \oplus \ell_p \approx (\ell_p \oplus \ell_p)_p = \ell_p,$$

where "$=$" means "is isometric to." (Why?)

Given a sequence of Banach spaces $X_1, X_2, \ldots$, we define the ℓ_p-*sum* of $X_1, X_2, \ldots$ to be the space of all sequences (x_n), with $x_n \in X_n$, for which $\|(x_n)\|^p = \sum_{n=1}^{\infty} \|x_n\|_{X_n}^p < \infty$, in case $p < \infty$, or $\|(x_n)\|_\infty = \sup_n \|x_n\|_{X_n} < \infty$, in case $p = \infty$, and we use the shorthand $(X_1 \oplus X_2 \oplus \cdots)_p$ to denote this new space. In brief, for any $1 \leq p \leq \infty$, we have

$$(X_1 \oplus X_2 \oplus \cdots)_p = \{(x_n) : x_n \in X_n \text{ and } (\|x_n\|)_{n=1}^{\infty} \in \ell_p\}.$$

The c_0-*sum* of spaces is defined in an entirely analogous fashion. In this case we write

$$(X_1 \oplus X_2 \oplus \cdots)_0 = \{(x_n) : x_n \in X_n \text{ and } (\|x_n\|)_{n=1}^{\infty} \in c_0\}.$$

Please note that in each case we have defined $(X_1 \oplus X_2 \oplus \cdots)_p$ to be a proper subspace of the formal sum $X_1 \oplus X_2 \oplus \cdots$. In particular, we will no longer be able to claim that $(X_1 \oplus X_2 \oplus \cdots)_p$ and $(X_1 \oplus X_2 \oplus \cdots)_q$ are isomorphic for $p \neq q$. Notice, for example, that $(\mathbb{R} \oplus \mathbb{R} \oplus \cdots)_p = \ell_p$.

It should also be pointed out that the order of the factors $X_1, X_2, \ldots$ in an ℓ_p sum does not matter; that is, if $\pi : \mathbb{N} \to \mathbb{N}$ is any permutation, then

$$(X_1 \oplus X_2 \oplus \cdots)_p = (X_{\pi(1)} \oplus X_{\pi(2)} \oplus \cdots)_p,$$

where "=" means "is isometric to." (Why?)

Although this may sound terribly complicated, all that we need for now is one very simple observation: We always have

$$(\ell_p \oplus \ell_p \oplus \cdots)_p = \ell_p \quad \text{and} \quad (c_0 \oplus c_0 \oplus \cdots)_0 = c_0,$$

for any $1 \leq p < \infty$. And why should this be true? The proof, in essence, is one sentence: $\mathbb{N}$ *can be written as the union of infinitely many, pairwise disjoint, infinite subsets.* (How does this help?)

Given this notation, the proof of Pełczyński's theorem is just a few lines.

Theorem 5.10. *Let X be one of the spaces ℓ_p, $1 \leq p < \infty$, or c_0. Then, every infinite-dimensional complemented subspace of X is isomorphic to X.*

Proof. For simplicity of notation, let's consider $X = \ell_p$ for some $1 \leq p < \infty$. The proof in case $X = c_0$ is identical.

If Y is an infinite-dimensional complemented subspace of ℓ_p, then we can write $\ell_p = Y \oplus Z$ for some Banach space Z. And, from Corollary 5.7, we can also write $Y = X_1 \oplus W$, where W is some Banach space and where $X_1 \approx \ell_p$. In brief, $Y \approx \ell_p \oplus W$. Thus,

$$\ell_p \oplus Y \approx \ell_p \oplus (\ell_p \oplus W) \approx (\ell_p \oplus \ell_p) \oplus W \approx \ell_p \oplus W \approx Y$$

since $\ell_p \oplus \ell_p \approx \ell_p$. Now for some prestidigitation:

$$\begin{aligned}
\ell_p \oplus Y &= (\ell_p \oplus \ell_p \oplus \cdots)_p \oplus Y \\
&\approx ((Y \oplus Z) \oplus (Y \oplus Z) \oplus \cdots)_p \oplus Y \\
&\approx (Z \oplus Z \oplus \cdots)_p \oplus (Y \oplus Y \oplus \cdots)_p \oplus Y \\
&\approx (Z \oplus Z \oplus \cdots)_p \oplus (Y \oplus Y \oplus \cdots)_p \\
&\approx ((Y \oplus Z) \oplus (Y \oplus Z) \oplus \cdots)_p = \ell_p.
\end{aligned}$$

Hence, $Y \approx \ell_p \oplus Y \approx \ell_p$. $\square$

Notes and Remarks

Essentially all of the results in this chapter can be attributed to Pełczyński [15, 113], who might fairly be called the father of modern Banach space theory. After the devastation of the Polish school during World War II, the study of linear functional analysis was slow to recover. Aleksander Pełczyński and Joram Lindenstrauss resurrected the lost arts in the late 1950s and early 1960s and went on to form new centers in Poland and in Israel, respectively. Along with Robert James in America, they founded a new school of Banach space theory. Needless to say, all three names will be cited frequently in these notes.

The various gliding hump arguments presented in this chapter are typical of the genre and, it seems, based on ideas that are both old and of some historical curiosity (for more on this, see [134] and [135]). The first known appearance of a gliding hump argument is due to Lebesgue in 1905 [90] (but see also [91]), who used it to show that if (f_n) is a weakly convergent sequence in L_1, then $(\|f_n\|_1)$ is bounded. A similar application of the gliding hump technique merits inclusion here:

Theorem 5.11 (The Uniform Boundedness Theorem). *Let* $(T_\alpha)_{\alpha \in A}$ *be a family of linear maps from a Banach space* X *into a normed space* Y. *If the family is pointwise bounded, then, in fact, it is uniformly bounded. That is, if*

$$\sup_{\alpha \in A} \|T_\alpha x\| < \infty,$$

for each $x \in X$, *then,*

$$\sup_{\alpha \in A} \|T_\alpha\| = \sup_{\alpha \in A} \sup_{\|x\|=1} \|T_\alpha x\| < \infty.$$

Proof. Suppose that $\sup_\alpha \|T_\alpha\| = \infty$. We will extract a sequence of operators (T_n) and construct a sequence of vectors (x_n) such that

(a) $\|x_n\| = 4^{-n}$, for all n, and
(b) $\|T_n x\| > n$, for all n, where $x = \sum_{n=1}^{\infty} x_n$.

To better understand the proof, consider

$$T_n x = T_n(x_1 + \cdots + x_{n-1}) + T_n x_n + T_n(x_{n+1} + \cdots).$$

The first term has its norm bounded by $M_{n-1} = \sup_\alpha \|T_\alpha(x_1 + \cdots + x_{n-1})\|$. We'll choose the central term so that

$$\|T_n x_n\| \approx \|T_n\| \|x_n\| >> M_{n-1}.$$

We'll control the last term by choosing $x_{n+1}, x_{n+2}, \ldots$ to satisfy

$$\left\| \sum_{k=n+1}^{\infty} x_k \right\| \leq \frac{1}{3} \|x_n\|.$$

It is now time for some details. Suppose that $x_1, \ldots, x_{n-1}$ and $T_1, \ldots, T_{n-1}$ have been chosen. Set

$$M_{n-1} = \sup_{\alpha \in A} \|T_\alpha(x_1 + \cdots + x_{n-1})\|.$$

Choose T_n so that

$$\|T_n\| > 3 \cdot 4^n (M_{n-1} + n).$$

Next, choose x_n to satisfy

$$\|x_n\| = 4^{-n} \quad \text{and} \quad \|T_n x_n\| > \frac{2}{3} \|T_n\| \|x_n\|.$$

This completes the inductive step.

It now follows from our construction that

$$\|T_n x_n\| > \frac{2}{3} \|T_n\| \|x_n\| > 2(M_{n-1} + n)$$

and

$$\|T_n(x_{n+1} + \cdots)\| \leq \|T_n\| \sum_{k=n+1}^{\infty} 4^{-k}$$

$$= \|T_n\| \cdot \frac{1}{3} \cdot 4^{-n}$$

$$= \frac{1}{3} \|T_n\| \|x_n\|$$

$$< \frac{1}{2} \|T_n x_n\|.$$

Thus,

$$\|T_n x\| \geq \|T_n x_n\| - \|T_n(x_1 + \cdots + x_{n-1})\| - \|T_n(x_{n+1} + \cdots)\|$$

$$> \frac{1}{2} \|T_n x_n\| - \|T_n(x_1 + \cdots + x_{n-1})\|$$

$$> (M_{n-1} + n) - M_{n-1} = n. \quad \square$$

The original proof of this theorem, due to Banach and Steinhaus in 1927 [10], is lost to us I'm sorry to report. As the story goes, Saks, the referee of their paper, suggested an alternate proof – the one that you see in most modern textbooks – using the Baire category theorem (see [32] and [151]). I am told by Joe Diestel that their original manuscript is thought to have been lost during the war. It's unlikely that we'll ever know their original method of proof, but it's a fair guess that their proof was very similar to the own given above. This is not based on idle conjecture: For one, the gliding hump technique was quite well known to Banach and Steinhuas and had already surfaced in their earlier work. More importantly, the technique was well known to many authors at the time; in particular, this is essentially the same proof given by Hausdorff in 1932 [67].

Curiously, this proof resurfaces every few years (in the *American Mathematical Monthly*, for example) under the label "a nontopological proof of the uniform boundedness theorem." See, for example, [51] and [68]. Apparently, the proof using Baire's theorem (itself an elusive result) is memorable, whereas the gliding hump proof (based solely on first principles) is not.

Theorem 5.8 (in the form of Exercise 5) is due to H. R. Pitt [120]. See Lindenstrauss and Tzafriri [94] for more on strictly singular operators and, more importantly, much more on the subspace structure of ℓ_p and c_0.

Corollary 5.7 and Theorem 5.10 might lead you to believe that the ℓ_p spaces have a rather simple subspace structure. Once we drop the word "complemented," however, the situation changes dramatically: *For $p \neq 2$, the space ℓ_p contains infinitely many mutually nonisomorphic subspaces* (cf., e.g., [94, 95]).

Pełczyński's decomposition method (Theorem 5.10) has one obvious practical disadvantage: It's virtually impossible to write down an explicit isomorphism! From this point of view, it's at best an existence proof.

We might paraphrase the conclusion of Pełczyński's theorem by saying that the spaces c_0 and ℓ_p, $1 \leq p < \infty$ are *prime* because they have no nontrivial "factors." It's also true that ℓ_∞ is prime, but the proof is substantially harder.

The heart of Pełczyński's method is that an infinite direct sum $Y \oplus Y \oplus \cdots$ is able to "swallow up" one more copy of Y. The decomposition method will generalize, under the right circumstances, leading to a Schröder–Bernstein-like theorem for Banach spaces. Please take note of the various ingredients used in the proof:

(i) Y embeds complementably in X, and X embeds complementably in Y,

(ii) $X = (X \oplus X \oplus \cdots)_X$, and

(iii) $X \oplus X \approx X$ (which may *fail* for certain spaces but actually follows from (ii) in this case).

There are several good survey articles on complemented subspaces and Schröder–Bernstein-like theorems for Banach spaces. Highly recommended are the articles by Casazza [20, 21, 22, 23] and Mascioni [99].

The Schröder–Bernstein theorem for Banach spaces reads: *If X is iso-morphic to a complemented subspace of Y, and if Y is isomorphic to a complemented subspace of X, then X and Y are isomorphic.* The theorem is known to hold under rather mild restrictions on X and Y; however, it has recently been shown to *fail* in general. Indeed, 1998 Fields medalist W. T. Gowers [59] constructed an example of a Banach space X that is isomorphic to its cube $X \oplus X \oplus X$ but not to its square $X \oplus X$.

Gowers put several long-standing open problems to rest in recent years. Another such problem asked whether ℓ_p and c_0 are the only prime Banach spaces. Gowers again solved this problem in the negative by providing an example of a Banach space that fails to be isomorphic to any of its proper subspaces [58, 60, 61].

Exercises

1. The sequence $x_1 = e_1$, $x_2 = e_2 + e_3$, $x_3 = e_4 + e_5 + e_6$, ... is a block basis of (e_n). Show "by hand" that (x_n) is not equivalent to (e_n) in ℓ_p, for $1 \leq p < \infty$, but that (x_n) is equivalent to (e_n) in c_0. What can you say about the block basis $y_n = x_n/n$?

2. Prove Corollary 5.6.

3. Prove Corollary 5.7.

4. If $T : c_0 \to \ell_p$, $1 \leq p < \infty$, is bounded and linear, show that $\|Te_n\|_p \to 0$.

5. Show that every bounded linear map $T : \ell_r \to \ell_p$, $1 \leq p < r < \infty$, or $T : c_0 \to \ell_p$, $1 \leq p < \infty$, is compact. [Hint: T is completely continuous.]

6. Show that every bounded linear map $T : c_0 \to \ell_r$ or $T : \ell_r \to c_0$, $1 \leq r < \infty$, is strictly singular.

7. Show that $(X \oplus Y)_1^* = (X^* \oplus Y^*)_\infty$ isometrically.

8. Prove that $\mathbb{N}$ can be written as the union of infinitely many pairwise disjoint infinite subsets.

9. Find a "natural" copy of $(\ell_p \oplus \ell_p \oplus \cdots)_p$ in $L_p(\mathbb{R})$.

10. If (X_n) is a sequence of Banach spaces, prove that $(X_1 \oplus X_2 \oplus \cdots)_p$ is a Banach space for any $1 \leq p \leq \infty$.

11. The proof of Theorem 5.10 requires that $((Y \oplus Z) \oplus (Y \oplus Z) \oplus \cdots)_p \approx (Z \oplus Z \oplus \cdots)_p \oplus (Y \oplus Y \oplus \cdots)_p$. Verify this claim.

12. Prove Theorem 5.10 in the case $X = c_0$.

Chapter 6

Special Properties of c_0, ℓ_1, and ℓ_∞

The spaces c_0, ℓ_1, ℓ_2, and ℓ_∞ play very special roles in Banach space theory. You're already familiar with the space ℓ_2 and its unique position as the sole Hilbert space in the family of ℓ_p spaces. We won't have much to say about ℓ_2 here. And by now you will have noticed that the space ℓ_∞ doesn't quite fit the pattern that we've established for the other ℓ_p spaces – for one, it's not separable and so doesn't have a basis. Nevertheless, we will be able to say a few meaningful things about ℓ_∞. The spaces c_0 and ℓ_1, on the other hand, play starring roles when it comes to questions involving bases in Banach spaces and in the whole isomorphic theory of Banach spaces for that matter. Unfortunately, we can't hope to even scratch the surface here. But at least a few interesting results are within our reach.

Throughout, (e_n) denotes the standard basis in c_0 or ℓ_1, and (e_n^*) denotes the associated sequence of coefficient functionals. As usual, (e_n) and (e_n^*) are really the same; we just consider them as elements of different spaces.

True Stories About ℓ_1

We begin with a "universal" property of ℓ_1 due to Banach and Mazur [8].

Theorem 6.1. *Every separable Banach space is a quotient of ℓ_1.*

Proof. Let X be a separable Banach space, and write $B_X^o = \{x : \|x\| < 1\}$ to denote the open unit ball in X. What we need to show is that there exists a linear map $Q : \ell_1 \to X$ such that $Q(B_{\ell_1}^o) = B_X^o$. What we'll actually do is construct a norm one map Q such that $\overline{Q(B_{\ell_1})} = B_X$. Just as in the proof of the Open Mapping theorem, this will then imply that $Q(B_{\ell_1}^o) = B_X^o$.

To begin, since X is separable, we can find a sequence (x_n) in B_X that is dense in B_X. We define $Q : \ell_1 \to X$ by setting $Qe_n = x_n$ and extending linearly. Thus, $\|Q\| \leq 1$ since $\|\sum_n a_n x_n\|_X \leq \sum_n |a_n| = \|\sum_n a_n e_n\|_1$. Clearly,

55

then, $Q(B_{\ell_1}) \subset B_X$ and $Q(B^o_{\ell_1}) \subset B^o_X$. Since (x_n) is dense in B_X, we also have $\overline{Q(B_{\ell_1})} = B_X$.

Now, given $x \in B^o_X$, we have $\|x\| < 1 - \varepsilon < 1$ for some $\varepsilon > 0$. Thus, $y = (1 - \varepsilon)^{-1}x \in B^o_X$, too. We will finish the proof by showing that $y = Qz$ for some $z \in \ell_1$ with $\|z\|_1 < (1 - \varepsilon)^{-1}$. That is, we will show that $(1 - \varepsilon)^{-1}x = y \in Q((1 - \varepsilon)^{-1}B^o_{\ell_1})$. Here we go: Given $0 < \delta < \varepsilon$, we have

$$y \in B^o_X \implies \exists n_1 \text{ such that } \|y - x_{n_1}\| < \delta$$
$$\implies y - x_{n_1} \in \delta B_X = \overline{\{\delta x_j : j \neq n_1\}}$$
$$\implies \exists n_2 \text{ such that } \|y - x_{n_1} - \delta x_{n_2}\| < \delta^2$$
$$\implies y - x_{n_1} - \delta x_{n_2} \in \delta^2 B_X = \overline{\{\delta^2 x_j : j \neq n_1\}}$$
$$\implies \exists n_3 \text{ such that } \|y - x_{n_1} - \delta x_{n_2} - \delta^2 x_{n_3}\| < \delta^3$$
$$\implies \text{ and so on } \ldots$$
$$\implies y = \sum_{j=1}^{\infty} \delta^{j-1}x_{n_j} = Q\left(\sum_{j=1}^{\infty} \delta^{j-1}e_{n_j}\right),$$

and $z = \sum_{j=1}^{\infty} \delta^{j-1}e_{n_j}$ has norm $(1 - \delta)^{-1} < (1 - \varepsilon)^{-1}$ in ℓ_1. $\square$

We will apply Banach and Mazur's result to show that ℓ_1 contains an uncomplemented subspace. To accomplish this, we first need some property of ℓ_1 that's not shared by every Banach space. One such property – and a rather dramatic example at that – is due to J. Schur from 1921 [133].

Theorem 6.2. *In ℓ_1, weak sequential convergence implies norm convergence.*

Before we prove Schur's theorem, let's note that a weakly convergent sequence is necessarily norm bounded: Given a weakly convergent sequence (x_n) in a normed space X, consider the sequence of functionals $(\hat{x}_n) \subset X^{**}$ acting on X^*. For each $f \in X^*$ the sequence $(f(x_n)) = (\hat{x}_n(f))$ is bounded since it's convergent. That is, the sequence $(\hat{x}_n)$ is *pointwise bounded* on X^*. By the Banach–Steinhaus theorem, this sequence must actually be uniformly bounded on B_{X^*}. In short,

$$\sup_{f \in B_{X^*}} \sup_n |\hat{x}_n(f)| = \sup_n \sup_{f \in B_{X^*}} |\hat{x}_n(f)| = \sup_n \|\hat{x}_n\| = \sup_n \|x_n\| < \infty.$$

Now it's easy to see that norm convergence always implies weak convergence, and so Schur's result states that the two notions of convergence coincide for *sequences* in ℓ_1. But, since the weak topology on any infinite-dimensional space is guaranteed to be weaker than the norm topology, there are *nets* in ℓ_1 that are weakly convergent but not norm convergent. We'll have more to say about this later.

Proof. (of Schur's theorem): Suppose that (x_n) is a bounded sequence in ℓ_1 such that $x_n \xrightarrow{w} 0$, but $\|x_n\|_1 \nrightarrow 0$. We'll arrive at a contradiction by constructing an $f \in \ell_\infty$ such that $f(x_n) \nrightarrow 0$.

We have, in particular, that $x_n(k) = e_k^*(x_n) \to 0$, as $n \to \infty$, for each k (where, at the risk of some confusion, we've written $x_n(k)$ for the k-th coordinate of x_n). That is, (x_n) tends "coordinatewise" to zero. By a standard gliding hump argument, we know that some subsequence of (x_n) is "almost disjoint." That is, after passing to a subsequence and relabeling, we may suppose that

(i) $\|x_n\|_1 \geq 5\varepsilon > 0$ for all n, and

(ii) for some increasing sequence of integers $1 \leq p_1 < q_1 < p_2 < q_2 < \cdots$ we have $\sum_{i < p_k} |x_k(i)| < \varepsilon$ and $\sum_{i > q_k} |x_k(i)| < \varepsilon$; hence, $\sum_{i=p_k}^{q_k} |x_k(i)| \geq 3\varepsilon$.

Now we define $f \in \ell_\infty$ by

$$f(i) = \begin{cases} \operatorname{sgn}(x_k(i)), & \text{if } p_k \leq i \leq q_k \text{ for some } k = 1, 2, \ldots, \\ 0, & \text{otherwise.} \end{cases}$$

Then $\|f\|_\infty \leq 1$ and, for any k,

$$|f(x_k)| \geq \sum_{i=p_k}^{q_k} |x_k(i)| - \sum_{i < p_k} |x_k(i)| - \sum_{i > q_k} |x_k(i)| \geq 3\varepsilon - 2\varepsilon = \varepsilon > 0,$$

which is the contradiction we needed. $\square$

Corollary 6.3. *Let* $Q : \ell_1 \to c_0$ *be a quotient map. Then* $\ker Q$ *is not complemented in* ℓ_1.

Proof. In the presence of a quotient map $Q : \ell_1 \to c_0$, we have that $\ell_1 / \ker Q \approx c_0$. If $\ker Q$ were complemented in ℓ_1, then $c_0 \approx \ell_1 / \ker Q$ would be isomorphic to a subspace of ℓ_1. That is, we could find an (into) isomorphism $T : c_0 \to \ell_1$. But $e_k \xrightarrow{w} 0$ in c_0, which easily implies that $T e_k \xrightarrow{w} 0$ in ℓ_1. But then, from Schur's theorem (Theorem 6.2), we would have $\|T e_k\|_1 \to 0$, which is impossible if T is an isomorphism. $\square$

The conclusion here is that there can be no (into) isomorphism $T : c_0 \to \ell_1$, a fact that we already know to be true for other reasons. All that's really new here is the *existence* of an onto map $Q : \ell_1 \to c_0$.

We could easily apply the same reasoning to any separable space that contains a weakly null normalized sequence. In particular, we could just as easily have used any ℓ_p, $1 < p < \infty$, in place of c_0. Thus ℓ_1 has uncountably many isomorphically distinct uncomplemented subspaces, which is a far cry from what happens in a Hilbert space!

Curiously, there is a measure of uniqueness in our last two results. Specifically, if X is a separable space not isomorphic to ℓ_1, then $\ker Q$ is isomorphically the same no matter what map Q from ℓ_1 onto X we might take. It's an open problem whether this property is an isomorphic characterization of ℓ_1 and ℓ_2. See [94, Theorem 2.f.8 and Problem 2.f.9].

While we're speaking of quotients, let's turn the tables and consider maps onto ℓ_1.

Theorem 6.4. *Let X be a Banach space. If there is a bounded, linear, onto map $T : X \to \ell_1$, then X contains a complemented subspace isomorphic to ℓ_1.*

Proof. From the Open Mapping theorem, $T(B_X) \supset \delta B_{\ell_1}$ for some $\delta > 0$. Consequently, we can find a bounded sequence (x_n) in X such that $Tx_n = e_n$. We first show that (x_n) is equivalent to (e_n). Now,

$$\left\| \sum_{n=1}^{\infty} a_n x_n \right\| \le \sum_{n=1}^{\infty} |a_n| \|x_n\| \le C \sum_{n=1}^{\infty} |a_n|$$

since (x_n) is bounded. On the other hand,

$$\|T\| \left\| \sum_{n=1}^{\infty} a_n x_n \right\| \ge \left\| \sum_{n=1}^{\infty} a_n e_n \right\|_1 = \sum_{n=1}^{\infty} |a_n|.$$

Thus, (x_n) is equivalent to (e_n). That is, the map $S : \ell_1 \to X$ defined by $Se_n = x_n$ and extended linearly to all of ℓ_1 is an isomorphism from ℓ_1 onto $[x_n]$.

The fact that $[x_n]$ is complemented is now easy. Indeed, notice that TS is the identity on ℓ_1. Thus, ST is a projection onto $[x_n]$. $\square$

The last result is a neat and tidy curiosity with little consequence, wouldn't you think? But to Pełczyński it's the starting point for a deep theorem (that we'll prove later in this chapter). And to Lindenstrauss it's but one among a wealth of results about liftings and extensions of operators along with a variety of interesting characterizations of c_0, ℓ_1, and ℓ_∞ (cf. e.g., [94, Section 2.f]). We will settle for just one such result that is both simple and timely.

Corollary 6.5. *Theorem 6.4 characterizes* ℓ_1, *up to isomorphism, among all separable infinite-dimensional Banach spaces.*

Proof. Suppose that Y is a separable infinite-dimensional Banach space with the property that whenever $T : X \to Y$ is onto, then X contains a complemented subspace isomorphic to Y. Then, in particular, taking $X = \ell_1$ and T a quotient map onto Y, we must have Y isomorphic to a complemented subspace of ℓ_1. But, from Pełczyński's theorem (Theorem 5.10), it follows that Y must be isomorphic to ℓ_1. $\square$

Our last special property of ℓ_1 in this section concerns what are sometimes called "almost isometries" or "small isomorphisms." Specifically, any space that contains an isomorphic copy of ℓ_1 contains an "almost isometric" copy of ℓ_1. The same is true for c_0, although we'll omit the (similar) proof. This result is due to James [74].

Theorem 6.6. *Let* $||| \cdot |||$ *be an equivalent norm on* ℓ_1. *For every* $\varepsilon > 0$ *there is a subspace Y of ℓ_1 such that* $(Y, ||| \cdot |||)$ *is* $(1 + \varepsilon)$-*isomorphic to* ℓ_1.

Proof. We may suppose that $\alpha |||x||| \leq \|x\|_1 \leq |||x|||$ for all $x \in \ell_1$, where $\alpha > 0$. Let $\varepsilon > 0$ and let (P_n) be the natural projections associated with the usual basis (e_n) of ℓ_1. For each n put

$$\lambda_n = \sup\{\|x\|_1 : |||x||| = 1, \ P_n x = 0\}.$$

Thus, $\|x\|_1 \leq \lambda_n |||x|||$ whenever $P_n x = 0$. Since $\ker P_n \supset \ker P_{n+1}$, the λ_n's decrease; hence, $\lambda_n \downarrow \lambda$ for some $\alpha \leq \lambda \leq 1$.

Fix n_0 such that $\lambda_{n_0} < \lambda(1 + \varepsilon)$. Since $\lambda_n > \lambda/(1 + \varepsilon)$ for all n, we can construct a block basic sequence (y_k) of (e_n) such that $|||y_k||| = 1$, $P_{n_0} y_k = 0$, and $\|y_k\| > \lambda/(1 + \varepsilon)$ for all k. Now, for every choice of scalars (a_k), we have $P_{n_0}(\sum_k a_k y_k) = 0$, and hence

$$\left|\left|\left| \sum_{k=1}^{\infty} a_k y_k \right|\right|\right| \geq \lambda_{n_0}^{-1} \left\| \sum_{k=1}^{\infty} a_k y_k \right\|_1$$

$$= \lambda_{n_0}^{-1} \sum_{k=1}^{\infty} |a_k| \|y_k\|_1$$

$$\geq \lambda_{n_0}^{-1}(1 + \varepsilon)^{-1} \lambda \sum_{k=1}^{\infty} |a_k|$$

$$\geq (1 + \varepsilon)^{-2} \sum_{k=1}^{\infty} |a_k|.$$

Of course, $\||\sum_k a_k y_k\|| \le \sum_k |a_k|$ follows from the triangle inequality. Consequently, (y_k) is $(1 + \varepsilon)^2$-equivalent to (e_n); that is, $[y_k]$ is $(1 + \varepsilon)^2$-isomorphic to ℓ_1. $\square$

The Secret Life of ℓ_∞

We begin by recalling a useful corollary of the Hahn–Banach theorem: For every vector x in a normed space X we have

$$\|x\| = \sup\{|f(x)| : f \in X^*, \|f\| = 1\}.$$

Indeed, for a given x, we can always find a norm one functional f in X^* such that $|f(x)| = \|x\|$. Of interest here is the fact that there are enough functionals in the unit sphere of X^* to recover the norm in X. The question for the day is whether any smaller subset of X^* will work.

Our first result offers a simple example of just such a reduction:

Lemma 6.7. *If X is a separable normed space, then we can find a sequence (f_n) in X^* such that*

$$\|x\| = \sup_n |f_n(x)| \qquad (6.1)$$

for every x in X. In particular, a countable family in X^* separates points in X.

Proof. Given (x_n) dense in X, choose norm one functionals (f_n) in X^* such that $f_n(x_n) = \|x_n\|$. It's not hard to see that (6.1) then holds. $\square$

Our interest in Lemma 6.7 is that the conclusion holds for some nonseparable spaces, too. For example, it easily holds for the *dual* of any separable space (if (y_n) is dense in the unit sphere of Y, where $X = Y^*$, check that $(\hat{y}_n)$ works). In particular, ℓ_∞ *has this property*. Indeed, if $e_n^*(x) = x_n$ has its usual meaning, then the sequence (e_n^*) fits the bill: $\|x\|_\infty = \sup_n |e_n^*(x)|$.

If some collection of functionals $\Gamma \subset X^*$ satisfies

$$\|x\|_X = \sup_{f \in \Gamma} |f(x)|,$$

we say that Γ is a *norming set* for X, or that Γ *norms* X. In this language, Lemma 6.7 says that separable spaces have countable norming sets. It's easy to see that *any space with this property is isometric to a subspace of ℓ_∞*. Indeed, if $\|x\| = \sup_n |f_n(x)|$, then define $T : X \to \ell_\infty$ by $Tx = (f_n(x))_{n=1}^\infty$.

Taken together, these observations produce an easy corollary.

Corollary 6.8. *A normed space X has a countable norming set if and only if it is isometric to a subspace of ℓ_∞.*

There are a couple of reasons for having taken this detour. For one, it's now pretty clear that every normed space X embeds isometrically into $\ell_\infty(\Gamma)$ for some Γ, where Γ depends on the size of a norming set in X^*. For another, these observations lend some insight into the nature of *complemented* subspaces of ℓ_∞. Here's why:

If $T : \ell_\infty \to \ell_\infty$ is continuous and linear, then $\ker T = \bigcap_{n=1}^{\infty} \ker(e_n^* \circ T)$.

In particular, any complemented subspace of ℓ_∞ is a countable intersection of kernels of continuous linear functionals. Our next task is to show that c_0 fails to be so writeable; that is, c_0 is not complemented in ℓ_∞. This result is due to Phillips [119] and Sobczyk [139]. The proof we'll give is due to Whitley [143].

Theorem 6.9. *c_0 is not complemented in ℓ_∞.*

Proof. We'll show that c_0 can't be written as $\bigcap_{n=1}^{\infty} \ker f_n$ for any sequence of functionals (f_n) on ℓ_∞. (We won't need to know very much about $(\ell_\infty)^*$ in order to do this.) What this amounts to, indirectly, is showing that ℓ_∞/c_0 has no countable norming set and hence can't be a subspace of ℓ_∞. The proof sidesteps consideration of the quotient space, though. We proceed in three steps:

1° There is an *uncountable* family (N_α) of *infinite* subsets of $\mathbb{N}$ such that $N_\alpha \cap N_\beta$ is *finite* for $\alpha \neq \beta$.

2° For each α, the function $x_\alpha = \chi_{N_\alpha}$ is in $\ell_\infty \setminus c_0$ (since N_α is infinite). If $f \in (\ell_\infty)^*$ with $c_0 \subset \ker f$, then $\{\alpha : f(x_\alpha) \neq 0\}$ is *countable*.

3° If $c_0 \subset \bigcap_{n=1}^{\infty} \ker f_n$, then some x_α is in $\bigcap_{n=1}^{\infty} \ker f_n$, too. Thus, $c_0 \neq \bigcap_{n=1}^{\infty} \ker f_n$.

First let's see how 3° follows from 1° and 2° (besides, it's easiest): From 2°, we would have that the set $\{\beta : f_n(x_\beta) \neq 0$ for some $n\}$ is countable. Thus, some α isn't in the set. That is, there is an α for which $f_n(x_\alpha) = 0$ for all n, and so 3° follows.

1° is due to Sierpinski [137]. Here's a clever proof (that Whitley attributes to one Arthur Kruse): Let (r_n) be a fixed enumeration of the rationals. For each *irrational* α, choose and fix a subsequence (r_{n_k}) converging to α and now define $N_\alpha = \{n_k : k = 1, 2, \ldots\}$. Each N_α is infinite, there are uncountably many N_α, and $N_\alpha \cap N_\beta$ has to be finite if $\alpha \neq \beta$! Very clever, no?

Finally we prove 2°. Suppose that $f \in (\ell_\infty)^*$ vanishes on c_0. For each n, let $A_n = \{\alpha : |f(x_\alpha)| \geq 1/n\}$. We'll show that A_n is finite. Suppose that $\alpha_1, \ldots, \alpha_k$ are distinct elements in A_n. If we define

$$x = \sum_{i=1}^{k} \operatorname{sgn}(f(x_{\alpha_i})) \cdot x_{\alpha_i},$$

then $f(x) \geq k/n$. But since f vanishes on c_0, we're allowed to change x in finitely many coordinates without affecting the value of $f(x)$. In particular, if we define $M_i = N_{\alpha_i} \setminus \bigcup_{j \neq i} N_{\alpha_j}$, then M_i differs from N_{α_i} by a finite set, and so setting

$$y = \sum_{i=1}^{k} \operatorname{sgn}(f(x_{\alpha_i})) \cdot \chi_{M_i}$$

gives $f(y) = f(x)$. But now that we've made the underlying sets disjoint, we also have $\|y\|_\infty \leq 1$, and hence

$$k \leq nf(x) = nf(y) \leq n\|f\|.$$

Thus, A_n is finite. $\square$

The functionals on ℓ_∞ that vanish on c_0 are precisely the elements of $(\ell_\infty/c_0)^*$. In fact, our proof of 2° actually computes the norm of x in the quotient space ℓ_∞/c_0.

Corollary 6.10. *c_0 is not isomorphic to a dual space.*

Proof. Suppose, to the contrary, that there is an isomorphism $T : c_0 \to X^*$ from c_0 onto the dual of a Banach space X. Then $T^{**} : \ell_\infty \to X^{***}$ is again an isomorphism and, further, $T^{**}|_{c_0} = T$. Now, from Dixmier's theorem (Theorem 2.2), there exists a projection $P : X^{***} \to X^{***}$ mapping X^{***} onto X^*. But then $T^{-1} P T^{**} : \ell_\infty \to \ell_\infty$ is a projection from ℓ_∞ onto c_0, contradicting Theorem 6.9. $\square$

As it happens, a closed infinite-dimensional subspace of ℓ_∞ is complemented precisely when it's isomorphic to ℓ_∞. We can prove half of this claim by showing that ℓ_∞ is "injective." This means that ℓ_∞ has a certain "Hahn–Banach" property. This, too, is due to Phillips [119].

Theorem 6.11. *Let Y be any subspace of a normed space X and suppose that $T : Y \to \ell_\infty$ is linear and continuous. Then T can be extended to a continuous linear map $S : X \to \ell_\infty$ with $\|S\| = \|T\|$.*

Proof. Clearly,

$$Ty = (e_n^*(Ty))_{n=1}^\infty = (y_n^*(y))_{n=1}^\infty$$

where $y_n^* = e_n^* \circ T \in Y^*$. If we let $x_n^* \in X^*$ be a Hahn–Banach extension of y_n^*, then $Sx = (x_n^*(x))_{n=1}^\infty$ does the trick:

$$\|Sx\|_\infty = \sup_n |x_n^*(x)|$$

$$\leq \left(\sup_n \|x_n^*\| \right) \cdot \|x\|$$

$$= \left(\sup_n \|y_n^*\| \right) \cdot \|x\| \leq \|T\| \|x\|.$$

Thus, $\|S\| \leq \|T\|$. That $\|S\| \geq \|T\|$ is obvious. $\square$

As an immediate corollary, notice that if ℓ_∞ is a closed subspace of some Banach space X, then ℓ_∞ is necessarily the range of a norm one projection on X. Indeed, the identity $I : \ell_\infty \to \ell_\infty$ (considered as a map into X) extends to a norm one map $P : X \to \ell_\infty$ (also considered as a map into X). Clearly, P is a projection. In short, ℓ_∞ *is norm one complemented in any superspace.* This fact is actually equivalent to the "extension property" of the previous theorem.

It's not hard to see that any space $\ell_\infty(\Gamma)$ has this same "Hahn–Banach extension property."

Confessions of c_0

Although c_0 is obviously not complemented in every superspace, it does share ℓ_∞'s "extension property" to a certain extent. In particular, c_0 is "separably injective." This observation is due to Sobczyk [139]; the short proof of the following result is due to Veech [142]. (If you don't know the Banach–Alaoglu theorem, don't panic. We'll review all of the necessary details in Chapter 12.)

Theorem 6.12. *Let Y be a subspace of a separable normed linear space X. If $T : Y \to c_0$ is linear and continuous, then T extends to a continuous linear map $S : X \to c_0$ with $\|S\| \leq 2\|T\|$.*

Proof. As before, let $y_n^* = e_n^* \circ T \in Y^*$ and let $x_n^* \in X^*$ be a Hahn–Banach extension of y_n^* with $\|x_n^*\| = \|y_n^*\| \leq \|T\|$. We would like to define $Sx = (x_n^*(x))_{n=1}^\infty$ as before, but we need to know that $Sx \in c_0$. That is, we want to replace (x_n^*) by a sequence of functionals that tend pointwise to 0 on X, but we don't want to tamper with their values on Y. What to do?

Since X is separable, we know that $B = \|T\| \cdot B_{X^*}$ is both compact and metrizable in the weak* topology. Let d be a metric on B that gives this topology. Now let $K = B \cap Y^\perp$ and notice that any weak* limit point of (x_n^*) must lie in K since $x_n^*(y) = y_n^*(y) \to 0$ for any $y \in Y$. That is, $d(x_n^*, K) \to 0$. This means that we can "perturb" each x_n^* by an element of K without affecting its value on Y.

Specifically, we choose a sequence (z_n^*) in K such that $d(x_n^*, z_n^*) \to 0$ and we define $Sx = (x_n^*(x) - z_n^*(x))_{n=1}^\infty$. Then $Sx \in c_0$, $Sy = (x_n^*(y)) = Ty$ for $y \in Y$, and $\|S\| \le 2\|T\|$. $\square$

As an immediate corollary we get that *c_0 is complemented by a projection of norm at most* 2 *in any* separable *space that contains it*. Our final result brings us full circle. This is the promised (and deep) result of Bessaga and Pełczyński [15] mentioned earlier.

Theorem 6.13. *Let Y be a Banach space. If Y^* contains a subspace isomorphic to c_0, then Y contains a complemented subspace isomorphic to ℓ_1. Thus, Y^* contains a subspace isomorphic to all of ℓ_∞. In particular, no separable dual space can contain an isomorphic copy of c_0 and, of course, c_0 itself is not isomorphic to any dual space.*

Proof. To prove the first assertion, we'll construct a map from Y onto ℓ_1. The second assertion will then follow easily.

To begin, let $T : c_0 \to Y^*$ be an isomorphism into. Then $T^* : Y^{**} \to \ell_1$ is onto, by Theorem 2.1. But we want a map on Y, so let's consider $S = T^*|_Y$. Now $\langle e_n, Sy \rangle = \langle y, Te_n \rangle$ for all $y \in Y$ and all n, and so we must have

$$Sy = (\langle y, Te_1 \rangle, \langle y, Te_2 \rangle, \ldots) = (Te_1(y), Te_2(y), \ldots),$$

where (e_n) is the usual basis for c_0.

Next we "pull back" a copy of (e_n^*), the basis for ℓ_1: Since T^* is onto, there exists a constant K and a sequence (y_n^{**}) in Y^{**} such that $\|y_n^{**}\| \le K$ and $T^*(y_n^{**}) = e_n^*$ for all n. We would prefer a sequence in Y with this property, for then we'd be finished. We'll settle for the next best thing: Since $K \cdot B_Y$ is weak* dense in $K \cdot B_{Y^{**}}$, we can find a sequence (y_n) in Y with $\|y_n\| \le K$ such that

$$|Te_n(y_n) - 1| < 1/n, \quad \text{and} \quad \sum_{i=1}^{n-1} |Te_i(y_n)| < 1/n.$$

(We've used the functionals $Te_1, \ldots, Te_n$ to specify a certain weak* neighborhood of y_n^{**} for each n. Of course, $y_n^{**}(Te_k) = T^* y_n^{**}(e_k) = e_n^*(e_k) = \delta_{n,k}$.) Thus, Sy_n has a large nth coordinate and very small entries in its first $n-1$

coordinates. What this means is that (Sy_n) has a subsequence equivalent to the usual basis in ℓ_1 (i.e., an "almost disjoint" subsequence) whose span is complemented in ℓ_1 by a projection P. In particular, after passing to a subsequence, we can find a constant M so that

$$\left\| \sum_n a_n y_n \right\| \leq K \sum_n |a_n| \leq K M \left\| \sum_n a_n S y_n \right\|_1 \leq K M \|S\| \left\| \sum_n a_n y_n \right\|.$$

That is, S is invertible on $[y_n] \approx \ell_1$ and hence $Q = S^{-1} P S$ is a projection from Y onto $[y_n]$. This completes the proof of the first assertion. In other words, we now have $Y \approx \ell_1 \oplus X$ for some $X \subset Y$. It follows that $Y^* \approx \ell_\infty \oplus X^*$, which finishes the proof. $\square$

Notes and Remarks

The proof of Theorem 6.1 is closely akin to Banach's proof of the Open Mapping theorem. It's quite clear that Banach understood completeness completely! Schur's theorem (Theorem 6.2) is an older gem that also appears in Banach's book [6]. It provides a classic example of a gliding hump argument. Our proofs of Theorem 6.4 and its corollary are largely borrowed from Diestel [33], who attributes these results to Lindenstrauss. Theorem 6.6 is sometimes called "James's nondistortion theorem." In brief, it says that ℓ_1 and c_0 are not "distortable" (a notion that we won't pursue further here). Embeddings into ℓ_∞ spaces, along the lines of Corollary 6.8, are semiclassical and can be traced back to Fréchet (see [49]). Theorem 6.11 and its companion Theorem 6.12 were among the first results in a frenzy of research during the late 1940s and early 1950s concerning injective spaces and Hahn–Banach-like extension properties. For more on the spaces c_0, ℓ_1, and ℓ_∞, see Day [29], Diestel [33], Jameson [76], and Lindenstrauss and Tzafriri [94].

Exercises

1. Find a weakly null normalized sequence in L_1 and conclude that L_1 and ℓ_1 are not isomorphic.

2. Complete the proof of Lemma 6.7.

3. Show that the conclusion of Lemma 6.7 also holds in case $X = Y^*$, where Y is separable.

4. If X is separable, show that X^* embeds isometrically in ℓ_∞.

5. Let X and Y be Banach spaces and let A be a bounded, linear map from X *onto* Y. Then, for every bounded linear map $T : \ell_1 \to Y$, there exists

a "lifting" $\tilde{T} : \ell_1 \to X$ such that $A\tilde{T} = T$. [Hint: Use the Open Mapping theorem to choose a bounded sequence (x_n) in X such that $Ax_n = Te_n$. The operator defined by $\tilde{T}e_n = x_n$ then does the trick.]

6. If H is any Hilbert space, prove that every bounded linear operator from H into ℓ_1 is compact.

7. Prove Theorem 6.6 for c_0.

8. Prove Theorem 6.11 for maps into $\ell_\infty(\Gamma)$. That is, prove that $\ell_\infty(\Gamma)$ is injective.

9. Prove that the following are equivalent for a normed space X:
   (i) If X is contained isometrically in a normed space Y, then X is complemented by a norm one projection on Y.
   (ii) If E is a subspace of a normed space F, then every bounded linear map $T : E \to X$ extends to a bounded linear map $S : F \to X$ with $\|S\| = \|T\|$.

   [Hint: We've already seen that (ii) implies (i). To prove that (i) implies (ii), first embed X isometrically in some $\ell_\infty(\Gamma)$ space and extend T (as a map into $\ell_\infty(\Gamma)$) using Theorem 6.11.]

Chapter 7
Bases and Duality

If (x_n) is a basis for X, you may have wondered whether the sequence of coordinate functionals (x_n^*) forms a basis for X^*. If we consider the pair (e_n) and (e_n^*), then it's easy to see that the answer is sometimes yes (in case $X = c_0$ and $X^* = \ell_1$, for example) and sometimes no (in case $X = \ell_1$ and $X^* = \ell_\infty$, for instance). As it happens, though, (x_n^*) is always a basic sequence; that is, it's always a basis for $[x_n^*]$. Let's see why:

Given $a_1, \ldots, a_n$ and $\varepsilon > 0$, choose $x = \sum_{i=1}^\infty c_i x_i$ with $\|x\| = 1$ such that

$$\sum_{i=1}^n a_i c_i = \left\langle x, \sum_{i=1}^n a_i x_i^* \right\rangle > \left\| \sum_{i=1}^n a_i x_i^* \right\| - \varepsilon.$$

Now if K is the basis constant of (x_i), then

$$K = K\|x\| \geq \left\| \sum_{i=1}^n c_i x_i \right\|.$$

Thus, for any $m > n$, we have

$$K \left\| \sum_{i=1}^m a_i x_i^* \right\| \geq \left\| \sum_{i=1}^n c_i x_i \right\| \left\| \sum_{i=1}^m a_i x_i^* \right\| \geq \left\langle \sum_{i=1}^n c_i x_i, \sum_{i=1}^m a_i x_i^* \right\rangle$$

$$= \sum_{i=1}^n a_i c_i > \left\| \sum_{i=1}^n a_i x_i^* \right\| - \varepsilon.$$

That is, since ε was arbitrary, (x_n^*) is a basic sequence with the same basis constant K.

It's also easy to see that if P_n is the canonical projection onto $[x_1, \ldots, x_n]$, then P_n^* is the canonical projection onto $[x_1^*, \ldots, x_n^*]$. Indeed, given $m, k > n$, we have

$$\left\langle \sum_{j=1}^{k} b_j x_j, \; P_n^* \left(\sum_{i=1}^{m} a_i x_i^* \right) \right\rangle = \left\langle P_n \left(\sum_{j=1}^{k} b_j x_j \right), \; \sum_{i=1}^{m} a_i x_i^* \right\rangle$$

$$= \left\langle \sum_{j=1}^{n} b_j x_j, \; \sum_{i=1}^{m} a_i x_i^* \right\rangle = \sum_{i=1}^{n} b_i a_i$$

$$= \left\langle \sum_{j=1}^{k} b_j x_j, \; \sum_{i=1}^{n} a_i x_i^* \right\rangle.$$

That is, $P_n^*(\sum_{i=1}^{m} a_i x_i^*) = \sum_{i=1}^{n} a_i x_i^*$. What's more, since $P_n x \to x$ for any x, it's easy to check that $P_n^* x^* \xrightarrow{w^*} x^*$ for any x^*, and hence span(x_n^*) is weak*-dense in X^*. If X is reflexive, then this already implies that (x_n^*) is a basis for X^* since the weak (= weak*) closure of a subspace coincides with its norm closure.

In any case, (x_n^*) *is a basis for* X^* *if and only if* $[x_n^*] = X^*$. Our next result supplies a test for this condition.

Theorem 7.1. *Let* (x_n) *be a basis for* X *with coordinate functionals* (x_n^*). *Then,* (x_n^*) *is a basis for* X^* *if and only if* $\lim_{n \to \infty} \|x^*\|_n = 0$, *for each* $x^* \in X^*$, *where* $\|x^*\|_n$ *is the norm of* x^* *restricted to the "tail space"* span$\{x_i : i > n\}$.

Proof. The forward implication is easy. If (x_n^*) is a basis for X^* and if $x^* = \sum_i a_i x_i^*$, then

$$\|x^*\|_n = \left\| \sum_{i=n+1}^{\infty} a_i x_i^* \right\| \to 0.$$

Now suppose that $\lim_{n \to \infty} \|x^*\|_n = 0$ for each $x^* \in X^*$. Given $x^* \in X^*$, first notice that the functional $x^* - \sum_{i=1}^{n} \langle x_i, x^* \rangle x_i^*$ vanishes on the span of $x_1, \ldots, x_n$. Thus, given $x = \sum_{i=1}^{\infty} c_i x_i$, we have

$$\left| \left\langle x, \; x^* - \sum_{i=1}^{n} \langle x_i, x^* \rangle x_i^* \right\rangle \right| = \left| \left\langle \sum_{i=n+1}^{\infty} c_i x_i, \; x^* \right\rangle \right| \le \|x^*\|_n \left\| \sum_{i=n+1}^{\infty} c_i x_i \right\|.$$

But

$$\left\| \sum_{i=n+1}^{\infty} c_i x_i \right\| \le \|x\| + \left\| \sum_{i=1}^{n} c_i x_i \right\| \le (1 + K)\|x\|,$$

where K is the basis constant for (x_i). Thus,

$$\left\| x^* - \sum_{i=1}^{n} \langle x_i, x^* \rangle x_i^* \right\| \leq (1 + K)\|x^*\|_n \to 0,$$

and hence (x_n^*) is a basis for X^*. $\square$

We say that a basis (x_n) is *shrinking* if $\lim_{n \to \infty} \|x^*\|_n = 0$ for each $x^* \in X^*$, as in our last result. That is, (x_n) is shrinking if (x_n^*) is a basis for X^*. The natural basis (e_n) is shrinking in c_0 and ℓ_p, $1 < p < \infty$, but not in ℓ_1.

If X has a shrinking basis, then X^{**} can be represented in terms of the basis too.

Theorem 7.2. *If (x_n) is a shrinking basis for a Banach space X, then X^{**} can be identified with the space of all sequences of scalars (a_n) for which $\sup_n \| \sum_{i=1}^{n} a_i x_i \| < \infty$. The correspondence is given by*

$$x^{**} \leftrightarrow (x^{**}(x_1^*), \, x^{**}(x_2^*), \, \ldots) = (a_1, a_2, \ldots), \tag{7.1}$$

and the norm of x^{**} is equivalent to (and, in case the basis constant is 1, actually equal to) $\sup_n \| \sum_{i=1}^{n} x^{**}(x_i^*) x_i \|$.

Proof. By renorming X, if necessary, we may assume that the basis constant of (x_n) is 1. This will simplify our arithmetic.

If we mimic a few of our previous calculations and use the fact that (x_n^*) is a basis with constant 1, it's not hard to see that $P_n^{**} x^{**} = \sum_{i=1}^{n} x^{**}(x_i^*) x_i$, $P_n^{**} x^{**} \xrightarrow{w^*} x^{**}$, and $\| P_n^{**} x^{**} \| \leq \| P_{n+1}^{**} x^{**} \| \leq \|x^{**}\|$. Thus, we have

$$\|x^{**}\| = \lim_{n \to \infty} \| P_n^{**} x^{**} \| = \sup_n \| P_n^{**} x^{**} \|.$$

Conversely, if (a_n) is a sequence of scalars such that $\sup_n \| \sum_{i=1}^{n} a_i x_i \| < \infty$, then any weak* limit point x^{**} of the bounded sequence $(\sum_{i=1}^{n} a_i x_i)_{n=1}^{\infty}$ satisfies $x^{**}(x_i^*) = a_i$ for all i. Thus, since (x_i^*) is a basis for X^*, the functional x^{**} is uniquely determined, and it follows that $x^{**} = \text{weak*-}\lim_{n \to \infty} \sum_{i=1}^{n} a_i x_i$. $\square$

Note that if (x_n) is a shrinking basis for X, then the canonical image of X in X^{**} corresponds to those sequences (a_n) for which $(\sum_{i=1}^{n} a_i x_i)$ is not only bounded but actually converges in norm. Indeed, from (7.1), we have

$$\hat{x} \leftrightarrow (\hat{x}(x_1^*), \, \hat{x}(x_2^*), \, \ldots) = (x_1^*(x), \, x_2^*(x), \, \ldots) = (a_1, a_2, \ldots) \tag{7.2}$$

for any $x = \sum_n a_n x_n$ in X.

Related to our first question is this: If X^* has a basis, does X have a basis? The answer turns out to be yes; in fact X can be shown to have a shrinking basis, but this is hard to see. A somewhat easier question is, If Y has a basis (y_n), what property of (y_n) will tell us whether Y is actually the dual of some other space (with a basis) X? Our next result supplies the answer.

A basis (x_n) for a Banach space X is called *boundedly complete* if $\sum_n a_n x_n$ converges whenever $\sup_n \| \sum_{i=1}^n a_i x_i \| < \infty$. Note that the usual basis (e_n) for ℓ_p, $1 \le p < \infty$, clearly has this property. On the other hand, (e_n) fails to be a boundedly complete basis for c_0.

Theorem 7.3. *If a basis (x_n) for a Banach space X is shrinking, then (x_n^*) is a boundedly complete basis for X^*. If a basis (x_n) for a Banach space X is boundedly complete, then X is isomorphic to the dual of a Banach space with a shrinking basis.*

Proof. Suppose that (x_n) is a shrinking basis for a Banach space X. We already know that (x_n^*) is a basis for X^*; we need to show that (x_n^*) is boundedly complete. So, suppose that (a_n) is a sequence of scalars such that $\| \sum_{i=1}^n a_i x_i^* \| \le M$ for all n.

To show that $\sum_n a_n x_n^*$ converges in X^*, it's enough (by Banach–Steinhaus) to show that the series $\sum_n a_n \langle x, x_n^* \rangle$ converges for each $x \in X$. For this, we check that the sum is Cauchy for each $x = \sum_n c_n x_n$. But,

$$
\left| \sum_{i=n}^m a_n \langle x, x_n^* \rangle \right| = \left| \left\langle x, \sum_{i=n}^m a_n x_n^* \right\rangle \right|
$$

$$
= \left| \left\langle \sum_{i=n}^m c_i x_i, \sum_{i=1}^n a_n x_n^* \right\rangle \right|
$$

$$
\le \left\| \sum_{i=1}^n a_n x_n^* \right\| \left\| \sum_{i=n}^m c_i x_i \right\|
$$

$$
\le M \left\| \sum_{i=n}^m c_i x_i \right\| \to 0 \quad \text{as} \quad m, n \to \infty.
$$

Now suppose that (x_n) is a boundedly complete basis for X. We'll prove that X is isomorphic to the dual of $Y = [x_n^*] \subset X^*$. To begin, let $J : X \to Y^*$ be defined by $(Jx)(y) = y(x)$. That is, $Jx = \hat{x}|_Y$, the functional $\hat{x}$ considered as a functional on Y. Note that $\|Jx\| \le \|x\|$ for free, and so J is at least continuous. The claim here is that J is an onto isomorphism.

To show that J is an isomorphism, it's enough to consider the action of J on span (x_n); that is, on vectors of the form $x = \sum_{i=1}^n a_i x_i$. Given such an x, choose $x^* \in X^*$ with $\|x^*\| = 1$ and $x^*(x) = \|x\|$. Then,

$$\|x\| = x^*(x) = x^*(P_n x) = (P_n^* x^*)(x) = (Jx)(P_n^* x^*)$$

because $P_n^* x^* \in Y$. And since $\|P_n^* x^*\| \leq K$, where K is the basis constant of (x_n^*), we have that $\|x\| \leq K \|Jx\|$. That is, J is bounded below and thus is an isomorphism.

To show that J is onto, observe that the sequence $(Jx_n) \subset Y^*$ is biorthogonal to (x_n^*). Thus, (Jx_n) is a basic sequence in Y^* with the same basis constant K as (x_n^*). In particular, given $y^* \in Y^*$, we have $\|\sum_{i=1}^n y^*(x_i^*) Jx_i\| \leq K \|y^*\|$ for all n. Thus, since (x_n) is boundedly complete, the series $\sum_n y^*(x_n^*)x_n$ converges to an element $x \in X$. It's not hard to see that $Jx = y^*$. $\square$

The notions of shrinking and boundedly complete, taken together, characterize reflexivity for spaces possessing a basis.

Theorem 7.4. *Let X be a Banach space with a basis (x_n). Then, X is reflexive if and only if (x_n) is both shrinking and boundedly complete.*

Proof. If (x_n) is shrinking, then, from Theorem 7.2, X^{**} corresponds to the collection of all sequences (a_n) for which $\sup_n \|\sum_{i=1}^n a_i x_i\| < \infty$ and $\widehat{X}$ corresponds to the collection of all sequences (a_n) for which $(\sum_{i=1}^n a_i x_i)$ converges in X. If (x_n) is also boundedly complete, it's clear that these two collections are identical. Comparing (7.1) and (7.2), we must have $\widehat{X} = X^{**}$.

Now suppose that X is reflexive. As we've seen, this already implies that (x_n^*) is a basis for X^*; that is, (x_n) is shrinking. It remains to show that (x_n) is boundedly complete.

Now since (x_n^*) is a basis for X^*, it has a corresponding sequence of coefficient functionals (x_n^{**}) in X^{**}. But it's easy to see that $x_n^{**} = \hat{x}_n$; that is, x_n^{**} is just x_n considered as an element of X^{**}. (Why?) Since X is reflexive, it's clear, then, that $(\hat{x}_n)$ is a basis for $\widehat{X} = X^{**}$. From Theorem 7.3, this means that (x_n^*) is a shrinking basis for X^*, which in turn means that $(\hat{x}_n)$ is a boundedly complete basis for $\widehat{X}$. Of course, all of this means that (x_n) itself is boundedly complete. $\square$

Notes and Remarks

All of the results in this chapter are due to R. C. James [72]; in fact, our presentation is largely based on his 1982 article in the *American Mathematical Monthly* [75].

Exercises

1. Let (e_n) denote the usual basis of ℓ_1. What is $[e_n^*]$ in ℓ_∞?

2. If (x_n) is a basis for X, show that $(\hat{x}_n) \subset X^{**}$ is a sequence biorthogonal to the basic sequence (x_n^*). In particular, $(\hat{x}_n)$ is a basic sequence with the same basis constant as (x_n).

3. If X has a basis (x_n) with canonical projections (P_n), then $P_n^* x^* \xrightarrow{w^*} x^*$ for every $x^* \in X^*$. In other words, (x_n^*) is a *weak* basis* for X^*.

4. If $x_n^{**} \xrightarrow{w^*} x^{**}$ in X^{**}, show that $\|x^{**}\| \leq \liminf_{n \to \infty} \|x_n^{**}\|$. If $\|x_n^{**}\| \leq \|x^{**}\|$ for all n, conclude that $\|x^{**}\| = \lim_{n \to \infty} \|x_n^{**}\|$.

5. Check directly that (e_n) is not a shrinking basis for ℓ_1 and not a boundedly complete basis for c_0.

6. Prove that $C[0, 1]^*$ is not separable and conclude that $C[0, 1]$ does not have a shrinking basis. [Hint: Consider the collection $\{\delta_t : 0 \leq t \leq 1\}$, where δ_t is the point mass at t; that is, the Borel measure defined by $\delta_t(A) = 1$ if $t \in A$ and $\delta_t(A) = 0$ otherwise.]

7. Prove that $C[0, 1]$ is not isomorphic to a dual space and conclude that $C[0, 1]$ does not have a boundedly complete basis. [Hint: Theorem 6.13 and Exercise 3 on page 46.]

8. If (x_n) is a boundedly complete basis for X, show that any block basis (y_k) of (x_n) is also boundedly complete.

Chapter 8
L_p Spaces

Throughout, $L_p = L_p[0, 1] = L_p([0, 1], \mathcal{B}, m)$, where $\mathcal{B}$ is the Borel σ-algebra and m is Lebesgue measure. As a collection of equivalence classes under equality a.e., it makes little difference whether we use the Borel σ-algebra or the larger σ-algebra of Lebesgue measurable sets. We will almost surely be glib about the use of "almost everywhere" and its relatives. Much of what we'll have to say holds in any space $L_p(\mu)$ but, as we've already pointed out, the spaces $L_p[0, 1]$ and ℓ_p are the most important for our purposes. If we have recourse to use other measure spaces, we will be careful to say so. Given any interval on the line, though, Lebesgue measure is always understood.

Basic Inequalities

We begin with a survey of important inequalities; most of these you already know.

First on the list has to be *Hölder's inequality*: Given $1 < p < \infty$, let q denote the conjugate exponent defined by $1/p + 1/q = 1$. If $f \in L_p$ and $g \in L_q$, then $fg \in L_1$ and $\|fg\|_1 \le \|f\|_p \|g\|_q$. Equality can only occur if $|f|^p$ and $|g|^q$ are proportional; that is, $\alpha |f|^p = \beta |g|^q$ for some constants α, β, not both zero.

In the case $p = 1$ and $q = \infty$, the inequality is nearly obvious. Equality in this case can only occur if $|g| = \|g\|_\infty$ almost everywhere on the support of f.

It follows that $\|f\|_p \le \|f\|_q$ whenever $1 \le p \le q \le \infty$ and $f \in L_q$. More generally, if $\mu(X) < \infty$, $1 \le p \le q \le \infty$, and $f \in L_q(\mu)$, then $\|f\|_p \le \mu(X)^{1/p - 1/q} \|f\|_q$. Equality can only occur if $|f| = c \cdot 1$; that is, $|f|$ must be a.e. constant. We'll give another proof of this fact shortly.

Two variants of Hölder's inequality are also useful. Each is proved by an appropriate application of the "regular" version of the inequality. The first is simply a fancier formulation that we'll call the *generalized Hölder inequality*: Let $1 \le p, q, r < \infty$ satisfy $1/r = 1/p + 1/q$. If $f \in L_p$ and $g \in L_q$,

73

then $fg \in L_r$ and $\|fg\|_r \le \|f\|_p \|g\|_q$. The second is called *Liapounov's inequality*: Let $1 \le p, q < \infty$ and $0 \le \lambda \le 1$. If $r = \lambda p + (1 - \lambda)q$, then $\|f\|_r^r \le \|f\|_p^{\lambda p} \|f\|_q^{(1-\lambda)q}$ for any $f \in L_p \cap L_q$. What this means is that the function $\log \|f\|_r^r$ is a convex function of r. We'll have more to say about this shortly.

Most authors use Hölder's inequality to deduce *Minkowski's inequality* (although it's possible to do this just the other way around): Given $1 < p < \infty$ and $f, g \in L_p$, we have $\|f + g\|_p \le \|f\|_p + \|g\|_p$. Equality can only occur if $\alpha f = \beta g$ for some *nonnegative* constants α, β not both zero.

Again, the inequality is nearly obvious in case $p = 1$ or $p = \infty$, although the case for equality changes. For $p = 1$, equality can only occur if $|f + g| = |f| + |g|$; that is, only if f and g have the same sign or, in brief, only if $fg \ge 0$. For $p = \infty$, the case for equality is harder to state; more or less, equality occurs only if f and g have the same sign on some set of positive measure where both functions "attain" their norms.

It's of interest to us that Minkowski's inequality *reverses* for $0 < p < 1$. That is, if f and g are *nonnegative* functions in L_p for $0 < p < 1$, then $\|f + g\|_p \ge \|f\|_p + \|g\|_p$. We'll sketch a proof of this later.

Everything we've had to say thus far, with one obvious exception, has little if anything to do with the underlying measure space. Nevertheless, it's worthwhile summarizing the differences in at least one important case: Recall that we can identify ℓ_p^n with $L_p(\mu)$, where μ is counting measure on $\{1, \dots, n\}$. Given $1 \le p < q < \infty$ and $x = (x_1, \dots, x_n) \in \mathbb{R}^n$ we have $\|x\|_q \le \|x\|_p \le n^{1/p - 1/q} \|x\|_q$. Equality in the first inequality forces $x = c \, e_k$ for some k; that is, x can have at most one nonzero term. Equality in the second inequality forces $|x| = c \, (1, \dots, 1)$; that is, $|x|$ must be constant. The first inequality is easy to prove using induction and elementary calculus; the second is just Hölder's inequality in this new setting.

Convex Functions and Jensen's Inequality

The fact that the various inequalities we've just discussed do not depend on the underlying measure space is easiest to explain by saying that the function $\varphi(x) = |x|^p$ is convex. Recall that a function $f : I \to \mathbb{R}$ defined on some interval I is said to be *convex* if

$$f(\lambda x + (1 - \lambda)y) \le \lambda f(x) + (1 - \lambda)f(y) \tag{8.1}$$

for all x, $y \in I$ and all $0 \le \lambda \le 1$. Geometrically, this says that the chord joining $(x, f(x))$ and $(y, f(y))$ lies above the graph of f. Equivalently, f is convex if and only if its *epigraph*

$$\text{epi}\,(f) = \{(x, y) : f(x) \leq y\}$$

is a convex subset of $I \times \mathbb{R}$.

Now if f is convex, and if $a < c < b$ in I, then

$$c = \frac{b-c}{b-a}\,a + \frac{c-a}{b-a}\,b,$$

and hence

$$f(c) \leq \frac{b-c}{b-a}\,f(a) + \frac{c-a}{b-a}\,f(b).$$

By juggling letters, we can rewrite this inequality as

$$\frac{f(c) - f(a)}{c - a} \leq \frac{f(b) - f(a)}{b - a} \leq \frac{f(b) - f(c)}{b - c}$$

whenever $a < c < b$. If we apply this same reasoning to any $x, y \in [c, d] \subset [a, b]$, we get

$$\frac{f(c) - f(a)}{c - a} \leq \frac{f(y) - f(x)}{y - x} \leq \frac{f(b) - f(d)}{b - d}. \tag{8.2}$$

It follows that f is "locally Lipschitz," and hence absolutely continuous, on each closed subinterval of I. Thus, f' exists a.e. and, from (8.2), f' is increasing. If f is twice differentiable, for example, we would have $f'' \geq 0$.

The last conclusion is the one we're really after: *If f is twice differentiable (everywhere) in I, then f is convex if and only if $f'' \geq 0$.* The backward implication is easy to fill in: If $f'' \geq 0$, then f' is increasing. Thus, from the mean value theorem:

$$f(\lambda x + (1 - \lambda)y) - f(x) \leq f'(\lambda x + (1 - \lambda)y)(1 - \lambda)(y - x)$$
$$\leq \left(\frac{1 - \lambda}{\lambda}\right)[f(y) - f(\lambda x + (1 - \lambda)y)],$$

which, after a bit of symbol manipulation, yields (8.1).

Given this simple criterion, there are plenty of convex functions around: Any "concave up" function from elementary calculus will do – in particular, e^x and $|x|^p$, where $1 \leq p < \infty$, are convex. A function f for which $-f$ is convex is called *concave*; thus, $|x|^p$, where $0 < p < 1$, and $\log x$ are concave.

As a quick application of convexity, let's prove Minkowski's inequality. Given $f, g \in L_p$, $1 \le p < \infty$, with f and g not both zero, we have

$$\frac{|f + g|}{\|f\|_p + \|g\|_p} \le \frac{|f(x)| + |g(x)|}{\|f\|_p + \|g\|_p} = \lambda \cdot \frac{|f|}{\|f\|_p} + (1 - \lambda) \cdot \frac{|g|}{\|g\|_p},$$

where $\lambda = \|f\|_p/(\|f\|_p + \|g\|_p)$ and $1 - \lambda = \|g\|_p/(\|f\|_p + \|g\|_p)$. Thus, since $|x|^p$ is convex,

$$\frac{|f + g|^p}{(\|f\|_p + \|g\|_p)^p} \le \lambda \cdot \frac{|f|^p}{\|f\|_p^p} + (1 - \lambda) \cdot \frac{|g|^p}{\|g\|_p^p}.$$

Integrating both sides leads to $\|f + g\|_p^p/(\|f\|_p + \|g\|_p)^p \le 1$, which is the same as Minkowski's inequality. Note that for $0 < p < 1$ and $f, g \ge 0$, the inequality reverses since $|x|^p$ is concave.

Returning to our geometric interpretation of convex functions, we note that if f is convex on I, and if $x_0 \in I$, then f dominates any of its tangent lines at x_0. That is, if m lies between the left- and right-hand derivatives of f at x_0, then

$$f(x) \ge f(x_0) + m(x - x_0). \tag{8.3}$$

In particular, if f is differentiable at x_0, then $f(x) \ge f(x_0) + f'(x_0)(x - x_0)$. For differentiable functions, this last inequality implies that f' is increasing and so characterizes convexity (using the same mean value theorem proof we gave a moment ago). Royden calls $y = f(x_0) + m(x - x_0)$ a *supporting line* for the graph of f at x_0.

Finally, let's return to our discussion of L_p spaces by proving a classical inequality due to Jensen in 1906 [79].

Theorem 8.1 (Jensen's Inequality). *Let φ be a convex function on $\mathbb{R}$ and let $f \in L_1$. If $\varphi(f) \in L_1$, then*

$$\int_0^1 \varphi(f(t)) \, dt \ge \varphi\left(\int_0^1 f(t) \, dt \right).$$

Proof. Let $\alpha = \int f$ and let $y = \varphi(\alpha) + m(x - \alpha)$ be a supporting line at α. Then,

$$\varphi(f(t)) \ge \varphi(\alpha) + m(f(t) - \alpha).$$

Integrating both sides of the inequality does the trick. $\square$

It follows that if $f \in L_1$, then $\int e^f \geq e^{\int f}$, for example. Closer to home, if $f \in L_p$, $1 \leq p < \infty$, then $\int |f|^p \geq (\int |f|)^p$. This is the alternate proof that we alluded to earlier. We'll see another application of Jensen's inequality shortly.

A Test for Disjointness

We next discuss a few inequalities that are intimately related to the lattice structure of L_p. For example, it's clear that if f and g are disjointly supported in L_p, then $|f + g|^p = |f - g|^p = |f|^p + |g|^p$ a.e., and hence $\|f + g\|_p^p = \|f - g\|_p^p = \|f\|_p^p + \|g\|_p^p$. It would be more interesting, though, to rephrase this fact as a "test" for disjointness of two functions in L_p. For this we need another elementary inequality:

Lemma 8.2. *Given* $1 \leq p < 2$ *and* $a, b \in \mathbb{R}$, *we have*

$$|a + b|^p + |a - b|^p \leq 2 \left(|a|^p + |b|^p \right).$$

For $p = 2$ *we get equality for any* a, b. *For* $2 < p < \infty$, *the inequality reverses. For* $p \neq 2$, *equality can only occur if* $ab = 0$.

Proof. We apply our basic inequalities in the two-dimensional L_p space ℓ_p^2. Specifically,

$$|a + b|^p + |a - b|^p \leq 2^{1-p/2} \left(|a + b|^2 + |a - b|^2 \right)^{p/2}$$

$$\text{(by Hölder)}$$

$$= 2^{1-p/2} \cdot 2^{p/2} \left(|a|^2 + |b|^2 \right)^{p/2}$$

$$\leq 2 \left(|a|^p + |b|^p \right) \quad \text{(since } p < 2\text{)}.$$

Equality in the first inequality forces $|a + b| = |a - b|$, or $ab = 0$. The same conclusion holds if we have equality in the second inequality. The proof in case $p > 2$ is essentially identical. $\square$

Theorem 8.3. *Let* $f, g \in L_p(\mu)$, $1 \leq p < \infty$, $p \neq 2$. *Then,* f *and* g *are disjointly supported if and only if*

$$\|f + g\|_p^p + \|f - g\|_p^p = 2 \left(\|f\|_p^p + \|g\|_p^p \right).$$

Proof. The function

$$h = |f + g|^p + |f - g|^p - 2 \left(|f|^p + |g|^p \right)$$

is of constant sign and integrates to give

$$\int h \, d\mu = \|f + g\|_p^p + \|f - g\|_p^p - 2\left(\|f\|_p^p + \|g\|_p^p\right).$$

Thus, f and g are disjointly supported if and only if $fg = 0$, if and only if $h = 0$, and if and only if $\int h \, d\mu = 0$. $\square$

Corollary 8.4. *If $T : L_p(\mu) \to L_p(\nu)$ is a linear isometry, $1 \le p < \infty$, $p \ne 2$, then T maps disjoint functions to disjoint functions.*

Proof. If T is an isometry, then $\|Tf\|_p = \|f\|_p$, $\|Tg\|_p = \|g\|_p$, and $\|Tf \pm Tg\|_p = \|f \pm g\|_p$ for any f and g in $L_p(\mu)$. Consequently, f and g are disjoint if and only if Tf and Tg are disjoint. $\square$

This last result tells us something about the subspace structure of L_p. We already know that L_p contains an isometric copy of ℓ_p spanned by any sequence of disjointly supported, nonzero functions, and now we know that this is the only way to get an *isometric* copy of ℓ_p inside L_p.

L_p contains other natural sublattices. For example, if we identify the space $L_p[0, 1/2]$ with the collection of all functions $f \in L_p$ supported in $[0, 1/2]$, for example, then $L_p[0, 1/2]$ is a complemented sublattice of L_p, which is isometric to all of L_p. The projection is obvious:

$$Pf = f \cdot \chi_{[0,1/2]},$$

and the isometry is nearly obvious: If we set $(Tf)(t) = 2^{1/p} f(2t)$ for $0 \le t \le 1/2$ and $f \in L_p$, then $Tf \in L_p[0, 1/2]$ and

$$\|Tf\|_p^p = \int_0^{1/2} |Tf(t)|^p \, dt = 2 \int_0^{1/2} |f(2t)|^p \, dt = \|f\|_p^p.$$

More generally, $L_p[a, b]$ is isometric to L_p, and from this it's a short leap to the conclusion that $L_p[0, \infty)$ and $L_p(\mathbb{R})$ are isometric to L_p. Indeed, if we partition $[0, 1]$ into countably many disjoint, nontrivial, intervals (I_n), then

$$L_p = (L_p(I_1) \oplus L_p(I_2) \oplus \cdots)_p$$
$$= (L_p[0, 1] \oplus L_p[1, 2] \oplus \cdots)_p = L_p[0, \infty).$$

Of course, L_p is also isometric to a sublattice of $L_p[0, \infty)$.

Conditional Expectation

Less obvious examples of L_p-subspaces of L_p are generated by sub-σ-algebras of the Borel σ-algebra $\mathcal{B}$. Given a sub-σ-algebra $\mathcal{B}_0$ of $\mathcal{B}$, we define the space $L_p(\mathcal{B}_0) = L_p([0, 1], \mathcal{B}_0, m)$ to be the collection of all

$\mathcal{B}_0$-measurable functions in L_p. Thus, $L_p(\mathcal{B}_0)$ is a closed subspace (and even a sublattice) of L_p. In fact, $L_p(\mathcal{B}_0)$ is complemented in L_p by a norm one projection. What's more, any subspace of L_p that is the range of a norm one projection has to be of this form. We'll prove the first claim in some detail; the second claim isn't terribly hard to prove, but we won't pursue it further here.

By way of a simple example, suppose that we partition $[0, 1]$ into finitely many disjoint, measurable sets $A_1, \ldots, A_n$, with $m(A_k) > 0$ for each k, and let $\mathcal{B}_0$ be the algebra generated by the sets A_k. The $\mathcal{B}_0$-measurable functions are precisely the simple functions of the form $\sum_{k=1}^{n} a_k \chi_{A_k}$, and the space $L_p(\mathcal{B}_0)$ is then isometric to ℓ_p^n.

It might also help matters if we wrote out the projection onto $L_p(\mathcal{B}_0)$. In this case it's given by

$$Pf = \sum_{k=1}^{n} \left(\frac{1}{m(A_k)} \int_{A_k} f \right) \chi_{A_k}.$$

Recall that $\|Pf\|_p \leq \|f\|_p$ (review the computation we used to show that the Haar system is a basis for L_p). Notice that Pf is constant on each A_k, of course, and that $\int_A Pf = \int_A f$ for every $\mathcal{B}_0$-measurable set A. Indeed,

$$\int_{A_j} Pf = \left(\frac{1}{m(A_j)} \int_{A_j} f \right) \int_{A_j} \chi_{A_j} = \int_{A_j} f.$$

This simple example is not far from the general situation: In some sense, f is "averaged" over all $\mathcal{B}_0$-measurable sets to arrive at its $\mathcal{B}_0$-measurable counterpart.

To build a norm one projection from L_p onto $L_p(\mathcal{B}_0)$, we need some way to map $\mathcal{B}$-measurable functions into $\mathcal{B}_0$-measurable functions without increasing their norms. The secret here is what is known as a conditional expectation operator. Given an integrable, $\mathcal{B}$-measurable function f, we define *the conditional expectation of* f *given* $\mathcal{B}_0$ to be the a.e. unique integrable, $\mathcal{B}_0$-measurable function g with the property that

$$\int_A f = \int_A g \quad \text{for every} \quad A \in \mathcal{B}_0, \tag{8.4}$$

and we write $g = \mathbb{E}(f \mid \mathcal{B}_0)$. The existence and uniqueness of g follow from the Radon–Nikodým theorem. That is, if we define $\mu(A) = \int_A f$ for $A \in \mathcal{B}_0$, and if we think of m as being defined only on $\mathcal{B}_0$, then μ is absolutely continuous with respect to m and g is the Radon–Nikodým derivative $d\mu/dm$.

It's often simpler to think of (8.4) as the defining statement, though; after all, g is completely determined, as an element of L_1, by the values $\int_A g$ as A

runs through $\mathcal{B}_0$. For example, it's easy to see how (8.4) forces $\mathbb{E}(f \mid \mathcal{B}_0)$ to be both linear and positive. Consequently, we must also have $|\mathbb{E}(f \mid \mathcal{B}_0)| \leq \mathbb{E}(|f| \mid \mathcal{B}_0)$, and now it's clear that conditional expectation is a contraction on both L_1 and L_∞. Indeed, if $f \in L_1$, then

$$\|\mathbb{E}(f \mid \mathcal{B}_0)\|_1 = \int_0^1 |\mathbb{E}(f \mid \mathcal{B}_0)| \leq \int_0^1 \mathbb{E}(|f| \mid \mathcal{B}_0) = \int_0^1 |f| = \|f\|_1.$$

And if $f \in L_\infty$, then

$$|\mathbb{E}(f \mid \mathcal{B}_0)| \leq \mathbb{E}(|f| \mid \mathcal{B}_0) \leq \mathbb{E}(\|f\|_\infty \mid \mathcal{B}_0) = \|f\|_\infty$$

since constant functions are $\mathcal{B}_0$-measurable. Thus, $\|\mathbb{E}(f \mid \mathcal{B}_0)\|_\infty \leq \|f\|_\infty$.

As it happens, a positive linear map that is a simultaneous contraction on L_1 and L_∞ is also a contraction on every L_p, $1 < p < \infty$. That this is so follows from an application of an ingenious interpolation scheme due to Marcinkiewicz in 1939 [98].

Theorem 8.5. *Suppose that* $T : L_1 \to L_1$ *is a linear map satisfying*

   (i) $Tf \geq 0$ *whenever* $f \geq 0$ *(i.e., T is positive),*
   (ii) $\|Tf\|_1 \leq \|f\|_1$ *for all* $f \in L_1$, *and*
   (iii) $Tf \in L_\infty$ *whenever* $f \in L_\infty$ *and* $\|Tf\|_\infty \leq \|f\|_\infty$.

Then, $Tf \in L_p$ whenever $f \in L_p$, for any $1 < p < \infty$, and $\|Tf\|_p \leq \|f\|_p$.

Proof. Since $|Tf| \leq T|f|$, it's enough to consider the case in which $f \geq 0$. So, let $1 < p < \infty$ and let $0 \leq f \in L_p$. Now, for each fixed $y > 0$, we can write $f = f_y + R_y$, where $f_y = f \wedge y$, and $R_y = f - f_y = (f - y)^+$, and where y is also used to denote the constant function $y \cdot 1 \in L_\infty$.

Since $f_y \leq y$ we have that $T(f_y) \leq T(y) \leq y$ from (iii), and so

$$Tf = T(f_y) + T(R_y) \leq y + T(R_y).$$

But $T(R_y) \geq 0$ because $R_y \geq 0$; hence,

$$(Tf - y)^+ \leq T(R_y) = T((f - y)^+).$$

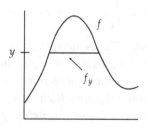

Next, integration gives

$$\int (Tf - y)^+ \le \int T((f - y)^+) = \|T((f - y)^+)\|_1$$

$$\le \|(f - y)^+\|_1 = \int (f - y)^+.$$

The rest of the proof consists in "upgrading" this inequality to an estimate involving the L_p norms of Tf and f.

We define auxiliary functions ρ_y and τ_y by

$$\rho_y(x) = \begin{cases} 1, & \text{if } f(x) - y > 0, \\ 0, & \text{otherwise,} \end{cases} \quad \text{and}$$

$$\tau_y(x) = \begin{cases} 1, & \text{if } (Tf)(x) - y > 0, \\ 0, & \text{otherwise.} \end{cases}$$

Then, $(f - y)^+ = (f - y)\rho_y$ and $(Tf - y)^+ = (Tf - y)\tau_y$. In this notation our last estimate says that

$$\int_0^1 (Tf - y)(x)\tau_y(x)\,dx \le \int_0^1 (f - y)(x)\rho_y(x)\,dx$$

for every $y > 0$. Consequently,

$$\int_0^\infty y^{p-2} \left(\int_0^1 (Tf(x) - y)\tau_y(x)\,dx \right) dy$$

$$\le \int_0^\infty y^{p-2} \left(\int_0^1 (f(x) - y)\rho_y(x)\,dx \right) dy.$$

Since all the functions involved are nonnegative, Fubini's theorem yields

$$\int_0^\infty y^{p-2} \left(\int_0^1 (f(x) - y)\rho_y(x)\,dx \right) dy$$

$$= \int_0^1 \int_0^\infty y^{p-2} (f(x) - y)\rho_y(x)\,dy\,dx$$

$$= \int_0^1 \int_0^{f(x)} y^{p-2} (f(x) - y)\,dy\,dx$$

$$= \left(\frac{1}{p-1} - \frac{1}{p} \right) \int_0^1 f(x)^p\,dx,$$

and, similarly,

$$\int_0^\infty y^{p-2} \left(\int_0^1 (Tf(x) - y)\tau_y(x)\,dx \right) dy$$

$$= \left(\frac{1}{p-1} - \frac{1}{p} \right) \int_0^1 Tf(x)^p\,dx.$$

Thus, $Tf \in L_p$ and $\|Tf\|_p \leq \|f\|_p$. □

Corollary 8.6. *Conditional expectation is a simultaneous contraction on every L_p.*

Some authors prove Corollary 8.6 by first proving a version of Jensen's inequality valid for conditional expectations; in particular, it can be shown that $|\mathbb{E}(f \mid \mathcal{B}_0)|^p \leq \mathbb{E}(|f|^p \mid \mathcal{B}_0)$. One proof of this fact first reduces to the case of nonnegative simple functions, where the inequality is more or less immediate, and then appeals to a "monotone convergence" theorem for conditional expectations (recall that conditional expectation is monotone). In any event, integration then yields $\|\mathbb{E}(f \mid \mathcal{B}_0)\|_p \leq \|f\|_p$.

There is an another approach to conditional expectation that warrants at least a brief discussion. In this approach, one begins by defining conditional expectation for L_2 functions. Specifically, we define $\mathbb{E}(\cdot \mid \mathcal{B}_0)$ to be the orthogonal projection from L_2 onto $L_2(\mathcal{B}_0)$. Since L_2 is dense in L_1, we can extend this definition to L_1 functions by taking monotone limits: Since each $0 \leq f \in L_1$ is the limit in L_1-norm of an increasing sequence (φ_n) of bounded (hence L_2) simple functions, we can take $\mathbb{E}(f \mid \mathcal{B}_0) = \lim_{n \to \infty} \mathbb{E}(\varphi_n \mid \mathcal{B}_0)$.

The savings in using Marcinkiewicz interpolation is clear: We have been spared the tedium of the typical "limit of simple functions" calculation. Instead we cut to the heart of the matter: Each $f \in L_p$ can be written as the sum of an L_∞ function and an L_1 function; namely, $f = f_y + R_y$. Marcinkiewicz then estimates the L_∞ norm of $\mathbb{E}(f_y \mid \mathcal{B}_0)$ and the L_1 norm of $\mathbb{E}(R_y \mid \mathcal{B}_0)$.

Finally, we should at least state the converse due to Douglas [38] (for $p = 1$) and Ando [4] (for $1 < p < \infty$).

Theorem 8.7. *Let $1 \leq p < \infty$, and let $P : L_p \to L_p$ be a positive, linear, contractive projection with $P(1) = 1$. Then there exists a sub-σ-algebra $\mathcal{B}_0$ such that $Pf = \mathbb{E}(f \mid \mathcal{B}_0)$ for every $f \in L_p$.*

Notes and Remarks

An excellent source for inequalities both big and small, including a discussion of convex functions and Jensen's inequality (Theorem 8.1), is the classic book *Inequalities* by Hardy, Littlewood, and Pólya [64]. The results on disjointly supported functions in L_p and their preservation under isometries were (essentially) known to Banach and form the basis for a more general result due to Lamperti [89]. Loosely stated, here is Lamperti's result for isometries on L_p.

Theorem 8.8. *Let* $1 \leq p < \infty$, $p \neq 2$, *and let* T *be a linear isometry on* $L_p[0, 1]$. *Then, there is a Borel measurable map* φ *from* $[0, 1]$ *onto* (*almost all of*) $[0, 1]$ *and a norm one* $h \in L_p$ *such that*

$$Tf = h \cdot (f \circ \varphi).$$

The function $h = T(1)$ *is uniquely determined* (*a.e.*), *and* φ *is uniquely determined* (*a.e.*) *on the support of* h. *Moreover, for any Borel set* E *we have*

$$m(E) = \int_{\varphi^{-1}(E)} |h|^p.$$

Lamperti's full result handles isometries from $L_p(\mu)$ into $L_p(\nu)$; details can be found in Royden [128], as can much of the material in the first three sections. Conditional expectation operators on L_p are by now part of the "folklore" of Banach space theory but can be traced to a rash of papers from the 1950s and 1960s (almost all of which appeared in the *Pacific Journal of Mathematics*), probably beginning with Moy [106] and certainly culminating in Ando [4]. Marcinkiewicz's theorem (Theorem 8.5) is a classic and can be found in any number of books; see, for example, [12]. The proof given here is taken from my notes for a course offered by D. J. H. Garling at The Ohio State University in the late 1970s [54].

Exercises

1. Let $1 \leq p, q, r < \infty$ satisfy $1/r = 1/p + 1/q$. If $f \in L_p(\mu)$ and $g \in L_q(\mu)$, show that $fg \in L_r(\mu)$ and $\|fg\|_r \leq \|f\|_p \|g\|_q$.

2. Let $1 \leq p, q < \infty$ and $0 \leq \lambda \leq 1$. If $r = \lambda p + (1 - \lambda)q$, show that $\|f\|_r^r \leq \|f\|_p^{\lambda p} \|f\|_q^{(1-\lambda)q}$ for any $f \in L_p(\mu) \cap L_q(\mu)$.

3. Given $1 < p < \infty$ and $f, g \in L_p(\mu)$ use Hölder's inequality to prove that $\|f + g\|_p \leq \|f\|_p + \|g\|_p$. Further, show that equality can only occur if $\alpha f = \beta g$ for some nonnegative constants α, β not both zero.

4. For $p = 1$, show that equality in Minkowski's inequality for $L_1(\mu)$ can only occur if $|f + g| = |f| + |g|$; that is, only if f and g have the same sign. Find a necessary and sufficient condition for equality when $p = \infty$.

5. Given $1 \leq p < q < \infty$ and $x \in \mathbb{R}^n$, prove that $\|x\|_q \leq \|x\|_p \leq n^{1/p - 1/q} \|x\|_q$. Further, show that equality in the first inequality forces $x = c \, e_k$ for some k, whereas equality in the second inequality forces $|x|$ to be constant.

6. Let $f : I \to \mathbb{R}$ be convex and let $x_0 \in I$.
   (a) Show that $f_l'(x_0) \leq f_r'(x_0)$, where f_l' (resp., f_r') is the left-hand (resp., right-hand) derivative of f at x_0.
   (b) If m lies between the left- and right-hand derivative of f at x_0, prove that $f(x) \geq f(x_0) + m(x - x_0)$ for all $x \in I$.

7. Prove Lemma 8.2 in the case $2 < p < \infty$.

8. If $\mathcal{B}_0$ is a sub-σ-algebra of $\mathcal{B}$, prove that $\mathbb{E}(\cdot \mid \mathcal{B}_0)$ is both linear and positive and conclude that $|\mathbb{E}(f \mid \mathcal{B}_0)| \leq \mathbb{E}(|f| \mid \mathcal{B}_0)$ for every $f \in L_1$.

9. If $\mathcal{B}_0$ is a sub-σ-algebra of $\mathcal{B}$, prove that $\mathbb{E}(\cdot \mid \mathcal{B}_0)$ is a projection on every L_p. [Hint: Use (8.4) to compute $\mathbb{E}(\mathbb{E}(f \mid \mathcal{B}_0) \mid \mathcal{B}_0)$.]

10. Given $1 \leq p < \infty$, a sub-σ-algebra $\mathcal{B}_0$ of $\mathcal{B}$, and a nonnegative simple function $f \in L_p$, show that $|\mathbb{E}(f \mid \mathcal{B}_0)|^p \leq \mathbb{E}(|f|^p \mid \mathcal{B}_0)$.

11. Let $\mathcal{B}_0$ be a sub-σ-algebra of $\mathcal{B}$. Prove that $Pf = \mathbb{E}(f \mid \mathcal{B}_0)$ is the orthogonal projection from L_2 onto $L_2(\mathcal{B}_0)$. [Hint: Use (8.4) to show that $\int g [f - \mathbb{E}(f \mid \mathcal{B}_0)] = 0$ for every $\mathcal{B}_0$-measurable simple function g and from this conclude that $\mathbb{E}(f \mid \mathcal{B}_0)$ is orthogonal to $f - \mathbb{E}(f \mid \mathcal{B}_0)$.]

Chapter 9

L_p Spaces II

We've seen that L_p contains copies of ℓ_p and L_p, but does L_p contain ℓ_q or L_q for $p \neq q$? In this chapter we'll settle the question completely in the case $2 \leq p < \infty$. (We'll consider the remaining case(s) in the next chapter.) We begin, though, with a classical result showing that every L_p contains a (natural) copy of ℓ_2.

The Rademacher Functions

We begin by describing an orthonormal system of functions on $[0, 1]$ of great importance in both classical and modern analysis. The *Rademacher functions* (r_n) are defined by $r_n(t) = \text{sgn}(\sin(2^n \pi t))$.

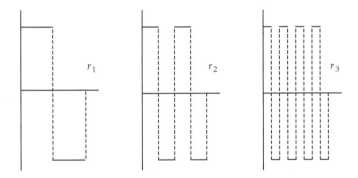

The Rademacher functions are related to the Haar system by $r_n = \sum_{k=2^{n-1}}^{2^n-1} h_k$, and from this it follows that (r_n) is an orthonormal sequence in L_2. Alternatively, notice that if $n < m$, then r_n is constant on each of the "periods" of r_m, and hence $\int r_n \cdot r_m = 0$. Since each r_n is mean zero, though, it's clear that (r_n) isn't complete (that is, the linear span of (r_n) isn't dense in L_2). Along the same lines, notice that if $n > 2$, then $r_1 r_2$ is constant on each of the "periods"

85

of r_n, and hence $\int r_n \cdot (r_1 r_2) = 0$. Thus, $r_1 r_2$ is orthogonal to the closed linear span of (r_n) in L_2.

If we were to work just a bit harder, we could show that $\int r_{n_1} \cdot r_{n_2} \cdots r_{n_k} = 0$ for any $n_1 < n_2 < \cdots < n_k$. The Rademacher functions, along with the functions of the form $r_{n_1} \cdot r_{n_2} \cdots r_{n_k}$ and the constant 1 function (sometimes labeled r_0), form the collection of *Walsh functions*, a complete orthonormal basis for L_2.

The Rademacher functions are important from a combinatorial point of view, too. Notice that as t ranges across $[0, 1]$, the vector $(r_1(t), \ldots, r_n(t))$ represents all possible choices of signs $(\varepsilon_1, \ldots, \varepsilon_n)$, where $\varepsilon_i = \pm 1$. Moreover, each particular choice $(\varepsilon_1, \ldots, \varepsilon_n)$ is equally likely, occurring on a set of measure 2^{-n}. We might say that $\sum_{k=1}^{n} a_k r_k(t) = \sum_{k=1}^{n} \pm a_k$ for some "random" choices of signs $\pm$. In particular,

$$\int_0^1 \left| \sum_{k=1}^n a_k r_k(t) \right| dt = 2^{-n} \sum_{\text{all } \varepsilon_i = \pm 1} \left| \sum_{k=1}^n \varepsilon_k a_k \right|$$

$$= \text{Average of } \left| \sum_{k=1}^n \pm a_k \right| \text{ over all choices } \pm.$$

An important classical inequality due to Khinchine in 1923 (below) tells us how to estimate this average. Note also that

$$\max_{0 \le t \le 1} \left| \sum_{k=1}^n a_k r_k(t) \right| = \sum_{k=1}^n |a_k|. \tag{9.1}$$

The Riesz–Fischer theorem tells us that $\sum_n a_n r_n$ converges in L_2 if and only if $\sum_n a_n^2 < \infty$. As it happens, though, $\sum_n a_n r_n$ converges in L_2 if and only if $\sum_n a_n r_n$ converges a.e. if and only if $\sum_n a_n^2 < \infty$. Thus, for example, the series $\sum \pm \frac{1}{n}$ converges for almost all choices of signs!

In probabilistic language, the Rademacher functions form a sequence of *independent Bernoulli random variables*. That is, the r_n are independent, identically distributed, mean zero random variables with $P\{r_n = +1\} = P\{r_n = -1\} = 1/2$. Also, if $f = \sum_{k=1}^n a_k r_k$, then $P\{f \in A\} = P\{-f \in A\}$; that is, f is a *symmetric* random variable.

There are a few good ways to see that the Rademacher functions are independent. Probably the easiest is this: Each $t \in [0, 1]$ has a binary decimal expansion $t = \sum_{n=1}^{\infty} \varepsilon_n(t)/2^n$, where $\varepsilon_n(t) = 0, 1$. The ε_n are clearly independent, and it's easy to check that $r_n(t) = 1 - 2\varepsilon_n(t)$. Alternatively, but along the same lines, we could identify $[0, 1]$ with the product space $\{-1, 1\}^{\mathbb{N}}$ by means of the map $t \mapsto (r_n(t))$, in which case r_n is the nth coordinate projection: $t \mapsto (\omega_k) \mapsto \omega_n = r_n(t)$. If we endow $\{-1, 1\}^{\mathbb{N}}$ with the product measure induced by taking counting measure in each coordinate, then $\{-1, 1\}^{\mathbb{N}}$ is measure-theoretically equivalent to $[0, 1]$. The point to this approach is that distinct coordinate projections in a product space are the canonical example of independent functions. No matter how we decide to say it, what we would need to check is

$$P\{r_n = 1, \ r_m = -1\} = 1/4 = P\{r_n = 1\} \cdot P\{r_m = -1\}, \quad \text{and}$$
$$P\{r_n = 1, \ r_m = 1\} = 1/4 = P\{r_n = 1\} \cdot P\{r_m = 1\}.$$

Induction takes care of the rest.

Because the Rademacher functions are independent it can be shown that

$$\int_0^1 f_1(r_1(t)) \cdots f_n(r_n(t)) \, dt = \left(\int_0^1 f_1(r_1(t)) \, dt \right) \cdots \left(\int_0^1 f_n(r_n(t)) \, dt \right)$$

for any n and any continuous functions $f_1, \ldots, f_n$. Some authors take this formula as the defining property of independence and we could do the same (it's the only consequence of independence that we'll need).

Khinchine's Inequality

It's a fact from classical Fourier analysis that any lacunary trigonometric sequence, such as $(\sin(2^n \pi t))$, satisfies

$$\left\| \sum_{k=1}^n a_k \sin(2^k \pi t) \right\|_p \approx \left(\sum_{k=1}^n |a_k|^2 \right)^{1/2}$$

for every $0 < p < \infty$. It's not too surprising that the Rademacher sequence shares this property.

Theorem 9.1 (Khinchine's Inequality). *Given* $0 < p < \infty$, *there exist constants* $0 < A_p, \ B_p < \infty$ *such that*

$$A_p \left(\sum_{k=1}^n |a_k|^2 \right)^{1/2} \leq \left\| \sum_{k=1}^n a_k r_k \right\|_p \leq B_p \left(\sum_{k=1}^n |a_k|^2 \right)^{1/2} \tag{9.2}$$

for all n and all scalars (a_k).

Proof. It should be pointed out that Khinchine's inequality works equally well for complex scalars (although our proof will use real scalars) but that it does not hold for $p = \infty$ (recall equation (9.1)).

Note that when $p = 2$ we can take $A_2 = 1 = B_2$. That is, we can write (9.2) as

$$A_p \left\| \sum_{k=1}^{n} a_k r_k \right\|_2 \leq \left\| \sum_{k=1}^{n} a_k r_k \right\|_p \leq B_p \left\| \sum_{k=1}^{n} a_k r_k \right\|_2. \tag{9.3}$$

Thus when $0 < p < 2$ we need only find A_p (and we take $B_p = B_2 = 1$), and when $2 < p < \infty$ we need only find B_p (and we take $A_p = A_2 = 1$).

As we'll see shortly, it's enough to establish the right-hand side of (9.2) for all $p \geq 2$. In fact, it's enough to consider the case where p *is an even integer*: $p = 2m$, $m = 1, 2, 3, \ldots$. In this case, we use the binomial theorem to compute

$$\int_0^1 \left| \sum_{k=1}^{n} a_k r_k(t) \right|^{2m} dt$$

$$= \sum_{k_1 + \cdots + k_n = 2m} \binom{2m}{k_1, k_2, \ldots, k_n} a_1^{k_1} a_2^{k_2} \cdots a_n^{k_n} \int_0^1 r_1^{k_1} r_2^{k_2} \cdots r_n^{k_n},$$

where we have used the *multinomial coefficients*

$$\binom{2m}{k_1, k_2, \ldots, k_n} = \frac{(2m)!}{k_1! \cdots k_n!}$$

and where the sum is over all nonnegative integers $k_1, k_2, \ldots, k_n$ that sum to $2m$. Now the integral in this formula is 0 unless every k_i is even, in which case it's 1. Thus,

$$\int_0^1 \left| \sum_{k=1}^{n} a_k r_k(t) \right|^{2m} dt$$

$$= \sum_{k_1 + \cdots + k_n = m} \binom{2m}{2k_1, 2k_2, \ldots, 2k_n} a_1^{2k_1} a_2^{2k_2} \cdots a_n^{2k_n}.$$

If we could replace the multinomial coefficient in this formula by $\binom{m}{k_1, k_2, \ldots, k_n}$, then we'd have $(\sum_{k=1}^{n} a_k^2)^m$ on the right-hand side. So, let's compare these two coefficients:

$$\frac{\binom{2m}{2k_1, 2k_2, \ldots, 2k_n}}{\binom{m}{k_1, k_2, \ldots, k_n}} = \frac{(2m)!}{(2k_1!) \cdots (2k_n!)} \cdot \frac{k_1! \cdots k_n!}{m!}$$

$$= \frac{(2m)(2m-1) \cdots (m+1)}{(2k_1) \cdots (k_1+1) \cdots (2k_n) \cdots (k_n+1)}$$

$$\leq \frac{2^m m^m}{2^{k_1} \cdots 2^{k_n}} = m^m.$$

Thus,

$$\int_0^1 \left| \sum_{k=1}^n a_k r_k(t) \right|^{2m} dt \leq m^m \cdot \sum_{k_1 + \cdots + k_n = m} \binom{m}{k_1, k_2, \ldots, k_n}$$

$$\times a_1^{2k_1} a_2^{2k_2} \cdots a_n^{2k_n} = m^m \left(\sum_{k=1}^n a_k^2 \right)^m,$$

or, $\| \sum_{k=1}^n a_k r_k \|_{2m} \leq m^{1/2} (\sum_{k=1}^n a_k^2)^{1/2}$. That is, $B_{2m} = m^{1/2}$ will work. In the general case we can take $B_p = (1 + p/2)^{1/2}$ because $p \leq 2(1 + [p/2]) \leq 2(1 + p/2)$. That is,

$$\left\| \sum_{k=1}^n a_k r_k \right\|_p \leq \left\| \sum_{k=1}^n a_k r_k \right\|_{2(1+[p/2])} \leq (1 + p/2)^{1/2} \left(\sum_{k=1}^n a_k^2 \right)^{1/2}.$$

Now when $0 < p < 2$, we use Liapounov's inequality: Choose $0 < \lambda < 1$ such that $2 = \lambda p + (1 - \lambda)4$. Then $f = \sum_{k=1}^n a_k r_k$ satisfies

$$\| f \|_2^2 \leq \| f \|_p^{p\lambda} \| f \|_4^{4(1-\lambda)}$$

$$\leq \| f \|_p^{p\lambda} (B_4 \| f \|_2)^{4(1-\lambda)} \leq 4^{(1-\lambda)} \| f \|_p^{p\lambda} \| f \|_2^{4(1-\lambda)}$$

since $B_4 = \sqrt{2}$. Thus, $\| f \|_p^{p\lambda} \geq 4^{(\lambda - 1)} \| f \|_2^{p\lambda}$ (because $2 - 4(1 - \lambda) = \lambda p$), and hence

$$\| f \|_p \geq 4^{-\left(\frac{1-\lambda}{p\lambda}\right)} \| f \|_2 = 4^{-\left(\frac{2-p}{2p}\right)} \| f \|_2$$

(because $\lambda = 2/(2 - p)$). That is, we can take $A_p = 2^{1-2/p}$. $\square$

Just for fun, here's another proof of the "hard" part of Khinchine's inequality. As before, the hard work is establishing the right-hand side of (9.2) for $p > 2$. Also, just as before, we may assume that p is an integer (but not necessarily even). In this case, $|f|^p \leq p! \, e^{|f|}$ and so it's enough to show that $\int e^{|f|} \leq C$ whenever $f = \sum_{k=1}^n a_k r_k$ and $\sum_{k=1}^n a_k^2 = 1$ (since (9.2) is homogeneous). Here goes:

$$\int e^{|f|} \leq 2 \int e^f \quad \text{(because } f \text{ is symmetric)}$$

$$= 2 \int \exp \left(\sum_{k=1}^n a_k r_k \right) = 2 \int \prod_{k=1}^n e^{a_k r_k}$$

$$= 2 \prod_{k=1}^{n} \int e^{a_k r_k} \quad \text{(by independence)}$$

$$= 2 \prod_{k=1}^{n} \cosh a_k.$$

But $\cosh x = \sum_{n=0}^{\infty} x^{2n}/(2n)! \le e^{x^2}$. Thus,

$$2 \prod_{k=1}^{n} \cosh a_k \le 2 \prod_{k=1}^{n} e^{a_k^2} = 2 \exp \left(\sum a_k^2 \right) = 2e.$$

And hence,

$$\|f\|_p \le (p!)^{1/p} \left\| e^{|f|} \right\|_p \le p \, (2e)^{1/p}.$$

That is, $B_p \le p \, (2e)^{1/p}$ in case p is an integer, and so $B_p \le (p+1)(2e)^{1/p}$ in general. Obviously, this value of B_p isn't quite as good as $(1 + p/2)^{1/2}$, but then this proof is shorter.

Khinchine's inequality yields an immediate corollary:

Corollary 9.2. *For any* $1 \le p < \infty$, *the Rademacher sequence* (r_n) *spans an isomorphic copy of* ℓ_2 *in* L_p. *If* $1 < p < \infty$, *then* $[r_n]$ *is even complemented in* L_p.

Proof. The first assertion is clear: In any L_p, $1 \le p < \infty$, the Rademacher sequence is equivalent to the usual basis of ℓ_2. Thus, the closed linear span of (r_n) in L_p is isomorphic to ℓ_2.

The second assertion is also easy. For $p = 2$, we have the orthogonal projection defined by

$$Pf = \sum_{n=1}^{\infty} \langle f, r_n \rangle r_n.$$

Of course

$$\|Pf\|_2 = \left(\sum_{n=1}^{\infty} |\langle f, r_n \rangle|^2 \right)^{1/2} \le \|f\|_2$$

for any $f \in L_2$. The claim here is that this same projection is bounded in every L_p for $1 < p < \infty$. For $p > 2$ this is immediate from Khinchine's inequality:

$$\|Pf\|_p \le B_p \left(\sum_{n=1}^{\infty} |\langle f, r_n \rangle|^2 \right)^{1/2} = B_p \|Pf\|_2 \le B_p \|f\|_2 \le B_p \|f\|_p.$$

For $1 < p < 2$ we use duality. Since (r_n) is orthonormal, the projection P is self-adjoint; that is,

$$\langle g, Pf \rangle = \sum_{n=1}^{\infty} \langle f, r_n \rangle \langle g, r_n \rangle = \langle Pg, f \rangle.$$

Thus, $P^* = P$ is a projection onto $[r_n]$ with norm at most B_q, where $q = p/(p-1)$ is the conjugate exponent (and $q > 2$ here). Since we're new to these sorts of computations, it couldn't hurt to say this another way. Notice that if $f \in L_p$ is fixed and $g \in L_q$ is arbitrary, we have

$$|\langle g, Pf \rangle| = |\langle Pg, f \rangle| \leq \|Pg\|_q \|f\|_p \leq B_q \|f\|_p \|g\|_q.$$

Thus, since g is arbitrary, we have $Pf \in L_p$ and $\|Pf\|_p \leq B_q \|f\|_p$. $\quad\square$

It's known that the closed linear span of the Rademacher functions is *not* complemented in L_1. Indeed, as we'll see, no complemented infinite-dimensional subspace of L_1 can be reflexive. In L_∞, the Rademacher functions span an isometric copy of ℓ_1 (recall equation (9.1)), which is known to be uncomplemented in L_∞. Nevertheless, L_∞ does contain an *isometric* copy of ℓ_2. (Indeed, ℓ_2 has a countable norming set and, hence, is isometric to a subspace of ℓ_∞.) This subspace is likewise uncomplemented.

Khinchine's inequality tells us even more:

Corollary 9.3. *For $1 \leq p < \infty$, $p \neq 2$, the spaces L_p and ℓ_p are not isomorphic.*

Indeed, as we've seen, ℓ_p, $1 \leq p < \infty$, $p \neq 2$, cannot contain an isomorphic copy of ℓ_2. Curiously, though, just as with ℓ_2 and L_2, the spaces ℓ_∞ and L_∞ are even isometric (but this is hard).

The Kadec–Pełczyński Theorem

We're now in a position to make a big dent in the problem raised at the beginning of this chapter. In particular, we'll completely settle the question of which L_q spaces embed into L_p for $p \geq 2$. The key is to generalize one of the properties of the Rademacher sequence: We will consider subspaces of L_p on which all of the various L_q norms for $1 \leq q < p$ are equivalent. The results in this section are due to Kadec and Pełczyński from 1962 [82]. Although this is one of the newer results that we'll see, it requires no machinery that was unknown to Banach; in this sense it qualifies as a classical result.

For $0 < \varepsilon < 1$ and $0 < p < \infty$, consider the following subset of L_p:

$$M(p, \varepsilon) = \{f \in L_p : m\{x : |f(x)| \geq \varepsilon \|f\|_p\} \geq \varepsilon\}.$$

Notice that if $\varepsilon_1 < \varepsilon_2$, then $M(p, \varepsilon_1) \supset M(p, \varepsilon_2)$. Also, $\bigcup_{\varepsilon>0} M(p, \varepsilon) = L_p$ since, for any nonzero $f \in L_p$, we have $m\{|f| \geq \varepsilon\} \nearrow m\{f \neq 0\} > 0$ as $\varepsilon \searrow 0$. In fact, any finite subset of L_p is contained in an $M(p, \varepsilon)$ for some $\varepsilon > 0$. Finally, note that if $f \in M(p, \varepsilon)$, then so is αf for any scalar α; in particular, $0 \in M(p, \varepsilon)$.

The elements of a given $M(p, \varepsilon)$ have to be relatively "flat." Notice, for example, that every element of $M(p, 1)$ is constant (a.e.). Indeed,

$$m\{|f| \geq \|f\|_p\} \geq 1 \Longrightarrow |f| = \|f\|_p \text{ a.e.}$$

Along similar lines, notice that $M(p, \varepsilon)$ doesn't contain the "spike" $f = \delta^{-1/p} \chi_{[0,\delta]}$ for any $0 < \delta < \varepsilon$. In this case, $\|f\|_p = 1$ while $m\{f \geq \varepsilon\} = \delta$. The following lemma puts this observation to good use.

Lemma 9.4. *For a subset A of L_p, the following are equivalent:*

  (i) *$A \subset M(p, \varepsilon)$ for some $\varepsilon > 0$.*
  (ii) *For each $0 < q < p$, there exists a constant $C_q < \infty$ such that $\|f\|_q \leq \|f\|_p \leq C_q \|f\|_q$ for all $f \in A$.*
  (iii) *For some $0 < q < p$, there exists a constant $C_q < \infty$ such that $\|f\|_q \leq \|f\|_p \leq C_q \|f\|_q$ for all $f \in A$.*

Proof. (i) $\Longrightarrow$ (ii): If $f \in A \subset M(p, \varepsilon)$, then

$$\int |f|^q = \int_{\{|f| \geq \varepsilon \|f\|_p\}} |f|^q + \int_{\{|f| < \varepsilon \|f\|_p\}} |f|^q$$
$$\geq \varepsilon^q \|f\|_p^q \, m\{|f| \geq \varepsilon \|f\|_p\}$$
$$\geq \varepsilon^{1+q} \|f\|_p^q.$$

That is, $\|f\|_q \geq \varepsilon^{(1+q)/q} \|f\|_p$.

(ii) $\Longrightarrow$ (iii) is clear, and so next is (iii) $\Longrightarrow$ (i): Suppose that $f \in A$ but that $f \notin M(p, \varepsilon)$ for some $\varepsilon > 0$. We will show that (iii) then implies $\varepsilon \geq \delta_0$ for some $\delta_0 > 0$ and hence that $A \subset M(p, \delta)$ for all $0 < \delta < \delta_0$.

Let $S = \{|f| \geq \varepsilon \|f\|_p\}$. Then, of course, $m(S) < \varepsilon$ because $f \notin M(p, \varepsilon)$. Next, write $1/q = 1/p + 1/r$ for some $r > 0$. Thus, $1 = q/p + q/r$ and so p/q and r/q are conjugate indices. Now let's estimate

$$\|f\|_q^q = \int_S |f|^q + \int_{S^c} |f|^q$$
$$\leq \left(\int_S |f|^p \right)^{q/p} \cdot m(S)^{q/r} + \varepsilon^q \|f\|_p^q$$
$$\leq (\varepsilon^{q/r} + \varepsilon^q) \|f\|_p^q.$$

Thus, from (iii), $\|f\|_p^q \leq C_q^q \|f\|_q^q \leq C_q^q(\varepsilon^{q/r} + \varepsilon^q)\|f\|_p^q$. Since $f \neq 0$, this means that ε must be bounded away from 0; that is, there exists some δ_0 such that $\varepsilon \geq \delta_0 > 0$. $\square$

If an entire subspace X of L_p is contained in some $M(p, \varepsilon)$, then the L_p and L_q topologies evidently coincide on X for every $0 < q < p$. In fact, it's even true that the L_p and L_0 topologies agree on X, where L_0 carries the topology of convergence in measure. Most important for our purposes is this: If $2 < p < \infty$, and if X is a closed subspace of L_p contained in some $M(p, \varepsilon)$, then the L_p and L_2 topologies coincide on X. That is, X must be isomorphic to a Hilbert space. How so? Well, if $\|f\|_2 \leq \|f\|_p \leq C\|f\|_2$ for all $f \in X$, then the inclusion map from X into L_2 is an isomorphism. Because every closed subspace of L_2 is isometric to a Hilbert space, X must be isomorphic to a Hilbert space.

Now if $f \in L_p$, $\|f\|_p = 1$ and $f \notin M(p, \varepsilon)$, then we can write $f = g + h$, where g is supported on a set of measure less than ε and where $\|h\|_p < \varepsilon$. Indeed,

$$g = f \cdot \chi_{\{|f| \geq \varepsilon\}} \quad \text{and} \quad h = f - g = f \cdot \chi_{\{|f| < \varepsilon\}}$$

do the trick. Note that the function g is a "spike" since $\|g\|_p > 1 - \varepsilon$ while $m(\operatorname{supp} g) < \varepsilon$.

Next, suppose that a subspace X of L_p fails to be entirely contained in $M(p, \varepsilon)$ for any $\varepsilon > 0$. Then, in particular, the set S_X of all norm one vectors in X isn't contained in any $M(p, \varepsilon)$. Thus, given a sequence of positive numbers $\varepsilon_n \to 0$, we can find a sequence of norm one vectors $f_n \in S_X$ such that $f_n \notin M(p, \varepsilon_n)$. Each f_n can be written $f_n = g_n + h_n$, where g_n is supported on a set of measure less than ε_n and where $\|h_n\|_p \leq \varepsilon_n$. That is, (f_n) is a small perturbation of the sequence of spikes (g_n). The claim here is that if $\varepsilon_n \to 0$ fast enough, then the seqence (g_n) is almost disjoint. If so, then (g_n), and hence also (f_n), will be equivalent to the usual basis in ℓ_p. This claim is worth isolating as a separate result.

Lemma 9.5. *Let (f_n) be a sequence of norm one functions in L_p, $1 \leq p < \infty$. If $m(\operatorname{supp} f_n) \to 0$, then some subsequence of (f_n) is equivalent to a disjointly supported sequence in L_p.*

Proof. The key observation here is that if $f \in L_p$ is fixed, then the measure $\mu(A) = \int_A |f|^p$ is absolutely continuous with respect to m. In particular, for each $\varepsilon > 0$ there is a $\delta > 0$ such that $\mu(A) < \varepsilon$ whenever $m(A) < \delta$.

Let $A_n = \text{supp } f_n$. Then $m(A_n) \to 0$. By induction, we can choose a subsequence of (f_n), which we again label (f_n), such that

$$\int_{A_{n+1}} \sum_{i=1}^{n} |f_i(t)|^p \, dt < 4^{-(n+1)p}$$

for all $n = 1, 2, \ldots$. Now let $B_n = A_n \setminus \bigcup_{i=n+1}^{\infty} A_i$ and let $g_n = f_n \cdot \chi_{B_n}$. Obviously, the sets (B_n) are pairwise disjoint and, hence, the sequence (g_n) is disjointly supported in L_p. Finally,

$$\|f_n - g_n\|_p^p = \int_{A_n \setminus B_n} |f_n(t)|^p \, dt \leq \sum_{i=n+1}^{\infty} \int_{A_i} |f_n(t)|^p \, dt$$

$$< \sum_{i=n+1}^{\infty} 4^{-ip} < 4^{-np},$$

and so $1 \geq \|g_n\|_p \geq 1 - 4^{-n} \geq 3/4$. Since (g_n) is a monotone basic sequence in L_p, the functionals (g_n^*) biorthogonal to (g_n) satisfy $\|g_n^*\|_q \leq 2/(3/4) = 8/3$ (where $1/p + 1/q = 1$). Thus, $\sum_n \|g_n^*\|_q \|f_n - g_n\|_p < (8/3)(1/3) < 1$. An appeal to the principle of small perturbations (Theorem 4.7) completes the proof. $\square$

It's easy to modify our last result to work for seminormalized sequences. In our particular application, this means that if we can find a sequence of norm one vectors $f_n \notin M(p, \varepsilon_n)$, where $\varepsilon_n \to 0$ fast enough, then (f_n) is a small perturbation of a sequence of "spikes" (g_n) with $\|g_n\|_p > 1 - \varepsilon_n$ and $m(\text{supp } g_n) < \varepsilon_n$. The seminormalized sequence (g_n) is in turn a small perturbation of a disjointly supported sequence in L_p. Thus, $[f_n]$ is isomorphic to ℓ_p and complemented in L_p. It's high time we summarized these observations:

Theorem 9.6. *Let* $2 < p < \infty$, *and let* X *be an infinite-dimensional closed subspace of* L_p. *Then, either*

  (i) X *is contained in* $M(p, \varepsilon)$ *for some* $\varepsilon > 0$, *in which case* X *is isomorphic to* ℓ_2 *and the* L_p *and* L_2 *topologies agree on* X, *or*
  (ii) X *contains a subspace that is isomorphic to* ℓ_p *and complemented in* L_p.

Corollary 9.7. *For* $2 < p < \infty$ *and* $1 \leq r < \infty$, $r \neq p, 2$, *no subspace of* L_p *is isomorphic to* L_r *or to* ℓ_r.

Proof. Since L_r contains an isometric copy of ℓ_r, it suffices to check that ℓ_r is not isomorphic to a subspace of L_p. But for $r \neq p$, 2, we know that ℓ_r is neither isomorphic to ℓ_2 nor does it contain a subspace isomorphic to ℓ_p. (Recall the discussion surrounding Theorem 5.8.) Thus, neither alternative of Theorem 9.6 is available. $\square$

As it happens, every copy of ℓ_2 in L_p, $p > 2$, is necessarily complemented. Curiously, this fails, in general, for $p < 2$.

Corollary 9.8. *Let X be a subspace of L_p, $2 < p < \infty$. If X is isomorphic to ℓ_2, then X is complemented in L_p.*

Proof. Since X cannot contain an isomorphic copy of ℓ_p, we must have $X \subset M(p, \varepsilon)$ for some $\varepsilon > 0$. Thus, there is a constant C such that $\|f\|_2 \leq \|f\|_p \leq C\|f\|_2$ for all $f \in X$. As we've seen, this means that X can also be considered as a subspace of L_2. As such, we can find an orthornormal sequence (φ_n) in L_2 such that X is the closed linear span of (φ_n), where the closure can be computed in either L_p or L_2 since the two topologies agree on X. Now let P be the orthogonal projection onto X; specifically, put

$$Pf = \sum_{n=1}^{\infty} \langle f, \varphi_n \rangle \varphi_n.$$

Just as we saw with the Rademacher functions, the projection P is also bounded as a map on L_p. Indeed, since $Pf \in X$, we have

$$\|Pf\|_p \leq C\|Pf\|_2 \leq C\|f\|_2 \leq C\|f\|_p. \quad \square$$

Corollary 9.9. *Let $2 < p < \infty$, and let X be an infinite-dimensional closed subspace of L_p. Then either X is isomorphic to ℓ_2 and complemented in L_p or X contains a subspace that is isomorphic to ℓ_p and complemented in L_p.*

By considering only *complemented* subspaces, we can use duality to transfer a modified version of these results to L_p for $1 < p < 2$. The principle at work is very general and well worth isolating:

Lemma 9.10. *Let Y be a closed subspace of a Banach space X. If Y is complemented in X, then Y^* is isomorphic to a complemented subspace of X^*.*

Proof. First recall that Y^* can be identified with a quotient of X^*. Specifically, $Y^* = X^*/Y^\perp$, isometrically. (Recall that if $i : Y \to X$ is inclusion, then $i^* : X^* \to Y^*$ is restriction and has kernel $Y^\perp$. Since i is an isometry into, i^* is a quotient map.)

Now if $P : X \to X$ is a projection with range Y, then $P^* : X^* \to X^*$ is a projection with kernel $Y^\perp$. As we've seen, P^* is indeed a projection. To see that ker $P^* = Y^\perp$, consider

$$P^*x^* = 0 \iff 0 = \langle x, P^*x^* \rangle = \langle Px, x^* \rangle \text{ for all } x \in X$$
$$\iff 0 = \langle y, x^* \rangle \text{ for all } y \in Y$$
$$\iff x^* \in Y^\perp.$$

Thus, the range of P^* can be identified, isomorphically, with $X^*/Y^\perp$, which we know to be Y^*, isometrically. That is, P^* is a projection onto an isomorphic copy of Y^*. $\square$

We will also need the following observation:

Lemma 9.11. *A closed subspace of a reflexive space is again reflexive.*

Proof. Let Y be a closed subspace of a reflexive Banach space X. Let $i : Y \to X$ denote inclusion, and let $j_Y : Y \to Y^{**}$ and $j_X : X \to X^{**}$ denote the canonical embeddings. Then $i^{**} : Y^{**} \to X^{**}$ is an isometry (into) that makes the following diagram commute:

$$
\begin{array}{ccc}
X & \xrightarrow{j_X} & X^{**} \\
i \uparrow & & \uparrow i^{**} \\
Y & \xrightarrow{j_Y} & Y^{**}.
\end{array}
$$

Thus, all four maps are isometries (into) and j_X is onto because X is reflexive. Next we compare ranges; for this we will need to compute annihilators (but in X^* and X^{**} only). Our calculations will be slightly less cumbersome if we occasionally ignore the formal inclusion i.

Now the range of i^{**} is $Y^{\perp\perp}$ (because the kernel of i^* is $Y^\perp$). Thus, Y is reflexive if and only if j_Y is onto if and only if $i^{**}(j_Y(Y)) = j_X(i(Y)) = Y^{\perp\perp}$. But, for any subset A of X^*, it's easy to check that $j_X(^\perp A) = A^\perp$ (since j_X is onto); thus, $j_X(^\perp(Y^\perp)) = Y^{\perp\perp}$. In other words, Y is reflexive if and only if $Y = i(Y) = {}^\perp(Y^\perp) = \overline{Y}$. $\square$

Corollary 9.12. *Let X be an infinite-dimensional complemented subspace of L_p, $1 < p < \infty$. Then either X is isomorphic to ℓ_2 or X contains a subspace that is isomorphic to ℓ_p and complemented in L_p.*

Proof. Of course, we already know this result when $2 \leq p < \infty$. So, suppose that $1 < p < 2$ and suppose that X is a complemented subspace of L_p. By Lemma 9.11 we know that X is reflexive, and by Lemma 9.10 we know that X^* is isomorphic to a complemented subspace of L_q, where $1/p + 1/q = 1$ and $q > 2$. Thus, either X^* is isomorphic to ℓ_2, or else X^* contains a complemented subspace isomorphic to ℓ_q. Now if X^* is isomorphic to ℓ_2, then surely $X^{**} = X$ is too. Finally, if X^* contains a complemented subspace that is isomorphic to ℓ_q, then $X^{**} = X$ contains a complemented subspace isomorphic to ℓ_p. $\square$

Corollary 9.13. *L_p is not isomorphic to L_q for any $p \neq q$.*

Now that we know L_p, $1 < p < \infty$, contains complemented subspaces isomorphic to ℓ_p and ℓ_2, it's possible to construct other, less apparent complemented subspaces. For example, it's not hard to see that $\ell_p \oplus \ell_2$ and $Z_p = (\ell_2 \oplus \ell_2 \oplus \cdots)_p$ are isomorphic to complemented subspaces of L_p. For $p \neq 2$, the spaces L_p, ℓ_p, ℓ_2, $\ell_p \oplus \ell_2$, and Z_p are isomorphically distinct (although that's not easy to see just now). In particular, L_p does contain complemented subspaces other than ℓ_p, ℓ_2, and L_p. In fact, L_p contains infinitely many isomorphically distinct complemented subspaces.

Notes and Remarks

An excellent source for information on the Rademacher functions is the 1932 paper by R. E. A. C. Paley [112]; see also Zygmund [150]. The "fancy" proof of Khinchine's inequality, which immediately follows the classical proof, is a minor modification of a proof presented in a course offered by Ben Garling at The Ohio State University in the late 1970s [54]. Stephen Dilworth has told me that this clever proof was shown to Garling by Simon Bernau. The constants A_p and B_p arising from our proof(s) of Khinchine's inequality are not the best possible; for example, it's known that $A_1 = 1/\sqrt{2}$ and $B_{2m} = ((2m-1)!!)^{1/2m}$, for $m = 1, 2, \ldots$, are best. See Szarek [141] and Haagerup [63]. Most of our applications of Khinchine's inequality to subspaces of L_p and, indeed, nearly all of the results from this chapter, are due to Kadec and Pełczyński [82], but much of the "flavor" of our presentation is borrowed from Garling's masterful interpretation in [54]. For more on the subspace structure of L_p see [95] and [81]. Lemma 9.11 is due to B. J. Pettis in 1938 [117] but is by now standard fare in most textbooks on functional analysis; see Megginson [100] for much more on reflexive subspaces.

Exercises

1. If $0 < \varepsilon_1 < \varepsilon_2 < 1$, show that $M(p, \varepsilon_1) \supset M(p, \varepsilon_2)$.

2. Prove that $\bigcup_{\varepsilon > 0} M(p, \varepsilon) = L_p$.

3. Prove Theorem 9.6.

4. Show that c_0 is not isomorphic to a subspace of L_p for $1 \leq p < \infty$. However, c_0 is isometric to a subspace of L_∞.

5. For $1 < p < \infty$, $p \neq 2$, prove that $\ell_p \oplus \ell_2$ and $Z_p = (\ell_2 \oplus \ell_2 \oplus \cdots)$ are isomorphic to complemented subspaces of L_p.

6. Let Y be a closed subspace of a Banach space X and let $i : Y \to X$ denote inclusion. Show that $i^* : X^* \to Y^*$ is restriction (defined by $i^*(f) = f|_Y$) and that $\ker i^* = Y^\perp$. Further, show that i^{**} is an isometry (into) with range $Y^{\perp\perp}$.

7. Let X be reflexive and let $j_X : X \to X^{**}$ denote the canonical embedding. Show that $j_X({}^\perp A) = A^\perp$ for any subset A of X^*. Is this true without the assumption that X is reflexive?

8. Fill in the details in the proof of Corollary 9.12. Specifically, suppose that X is a complemented subspace of L_p, $1 < p < 2$, and that X is not isomorphic to ℓ_2. Prove that X contains a complemented subspace isomorphic to ℓ_p.

9. Prove Corollary 9.13.

Chapter 10

L_p Spaces III

As pointed out earlier, the spaces L_p and ℓ_p exhaust the "isomorphic types" of $L_p(\mu)$ spaces. Thus, to better understand the isomorphic structure of $L_p(\mu)$ spaces, we might ask, as Banach did:

For $p \neq r$, can ℓ_r or L_r be isomorphically embedded into L_p?

We know quite a bit about this problem. We know that the answer is always yes for $r = 2$, and, in case $2 < p < \infty$, the Kadec–Pełczyński theorem (Corollary 9.7) tells us that $r = 2$ is the only possibility. In this chapter we'll prove the following statement:

If p and r live on opposite sides of 2, there can be no isomorphism from L_r or ℓ_r into L_p.

This leaves open only the cases $1 \leq r < p < 2$ and $1 \leq p < r < 2$. The first case can also be eliminated, as we'll see, but not the second.

Unconditional Convergence

We next introduce the notion of *unconditional convergence* of series. What follows are several plausible definitions.

We say that a series $\sum_n x_n$ in a Banach space X is

(a) *unordered convergent* if $\sum_n x_{\pi(n)}$ converges for every permutation (one-to-one, onto map) $\pi : \mathbb{N} \to \mathbb{N}$;

(b) *subseries convergent* if $\sum_k x_{n_k}$ converges for every subsequence (x_{n_k}) of (x_n);

(c) *"random signs" convergent* if $\sum_n \varepsilon_n x_n$ converges for any choice of signs $\varepsilon_n = \pm 1$;

(d) *bounded multiplier convergent* if $\sum_n a_n x_n$ converges whenever $|a_n| \leq 1$.

As we'll see, in a *complete* space all four notions are equivalent. Henceforth, we will say that $\sum_n x_n$ is *unconditionally convergent* if any one of these four conditions holds.

In $\mathbb{R}^n$, all four of these conditions are equivalent to $\sum_n \|x_n\| < \infty$; that is, all are equivalent to the absolute summability of $\sum_n x_n$. In an infinite-dimensional space, however, this is no longer the case. It is a deep result, due to Dvoretzky and Rogers in 1940 [45], that every infinite-dimensional space contains an unconditionally convergent series that fails to be absolutely summable.

We will briefly sketch the proof that (a) through (c) are equivalent (and leave condition (d) as an exercise), but first let's look at an example or two: If we fix an element $\sum_n a_n e_n$ in ℓ_p, $1 \le p < \infty$, then $\sum_n a_n e_n$ is unconditionally convergent (but not absolutely convergent, in general, unless $p = 1$). Why? Because $\sum_n a_n e_n$ converges in ℓ_p if and only if $\sum_n |a_n|^p < \infty$, which clearly allows us to change the signs of the a_n. For this reason, we sometimes say that ℓ_p has an *unconditional basis*. Similarly, if (e_n) is any orthonormal sequence in a Hilbert space H, then $\sum_n a_n e_n$ converges if and only if $\sum_n |a_n|^2 < \infty$ if and only if $\sum_n a_n e_n$ is unconditionally convergent.

Theorem 10.1. *Given a series $\sum x_n$ in a Banach space X, the following are equivalent:*

(i) *The series $\sum x_{\pi(n)}$ converges for every permutation π of $\mathbb{N}$.*
(ii) *The series $\sum x_{n_k}$ converges for every subsequence $n_1 < n_2 < n_3 < \cdots$.*
(iii) *The series $\sum \varepsilon_n x_n$ converges for every choice of signs $\varepsilon_n = \pm 1$.*
(iv) *For every $\varepsilon > 0$, there exists an N such that $\|\sum_{i \in A} x_i\| < \varepsilon$ for every finite subset A of $\mathbb{N}$ satisfying $\min A > N$.*

Proof. That (ii) and (iii) are equivalent is easy. The fact that (iv) implies (i) and (ii) is pretty clear since (iv) implies that each of the series in (i) and (ii) is Cauchy. The hard work comes in establishing (i), (ii) $\implies$ (iv). To this end, suppose that (iv) fails. Then, there exist an $\varepsilon > 0$ and finite subsets A_n of $\mathbb{N}$ satisfying $\max A_n < \min A_{n+1}$ and $\|\sum_{i \in A_n} x_i\| \ge \varepsilon$ for all n. But then, $A = \bigcup_n A_n$ defines a subsequence of $\mathbb{N}$ for which $\sum_{i \in A} x_i$ doesn't converge. Thus, (ii) fails. Lastly, we can find a permutation π of $\mathbb{N}$ such that π maps the interval $[\min A_n, \max A_n]$ onto itself in such a way that $\pi^{-1}(A_n) = B_n$ is an interval. Thus, $\|\sum_{i \in B_n} x_{\pi(i)}\| = \|\sum_{i \in A_n} x_i\|$. But then, $\sum x_{\pi(n)}$ doesn't converge, and so (i) also fails. $\square$

If $\sum x_n$ converges unconditionally to x, then condition (iv) implies that every rearrangement $\sum x_{\pi(n)}$ likewises converges to x. The same, of course, is no longer true of the sums $\sum \varepsilon_n x_n$. On the other hand, condition (iv) tells us that the set of all vectors of the form $\sum \varepsilon_n x_n$ is a *compact* subset of X. Indeed, from (iv), the map $f : \{-1, 1\}^{\mathbb{N}} \to X$ defined by $f((\varepsilon_n)) = \sum \varepsilon_n x_n$ is continuous. In particular, if $\sum x_n$ is unconditionally convergent, then

$$\sup_n \sup_{\varepsilon_i = \pm 1} \left\| \sum_{i=1}^{n} \varepsilon_i x_i \right\| < \infty. \tag{10.1}$$

It also follows from (iv) that if $\sum x_n$ is unconditionally convergent, then

(v) $\sum a_n x_n$ *converges for every bounded sequence of scalars* (a_n)

and the map $T : \ell_\infty \to X$ defined by $T((a_n)) = \sum a_n x_n$ is continuous (moreover, the restriction of T to c_0 is even compact). Since (v) clearly implies (iii), this means that condition (v) is another equivalent formulation of unconditional convergence. Because we have no immediate need for this particular condition, we will leave the details as an exercise.

Orlicz's Theorem

In terms of the Rademacher functions (or had you forgotten already?), we can write (10.1) another way: If $\sum_n x_n$ is unconditionally convergent, then

$$\sup_n \sup_{0 \le t \le 1} \left\| \sum_{i=1}^{n} r_i(t) x_i \right\| < \infty.$$

Armed with this observation we can make short work of an important theorem due to Orlicz in 1930 [111]. We begin with a useful calculation.

Proposition 10.2. *For any* $f_1, \ldots, f_n \in L_p$, $1 \le p < \infty$, *we have*

$$A_p \left\| \left(\sum_{i=1}^{n} |f_i|^2 \right)^{1/2} \right\|_p \le \left(\int_0^1 \left\| \sum_{i=1}^{n} r_i(s) f_i \right\|_p^p \, ds \right)^{1/p}$$

$$\le B_p \left\| \left(\sum_{i=1}^{n} |f_i|^2 \right)^{1/2} \right\|_p,$$

where $0 < A_p, B_p < \infty$ *are Khinchine constants.*

Proof. This is a simple matter of applying Fubini's theorem. First,

$$\left(\int_0^1 \left\| \sum_{i=1}^n r_i(s) f_i \right\|_p^p ds \right)^{1/p} = \left(\int_0^1 \int_0^1 \left| \sum_{i=1}^n r_i(s) f_i(t) \right|^p dt\, ds \right)^{1/p}$$

$$= \left(\int_0^1 \int_0^1 \left| \sum_{i=1}^n r_i(s) f_i(t) \right|^p ds\, dt \right)^{1/p}.$$

And now, from Khinchine's inequality,

$$\left(\int_0^1 \int_0^1 \left| \sum_{i=1}^n r_i(s) f_i(t) \right|^p ds\, dt \right)^{1/p}$$

$$\geq A_p \left(\int_0^1 \left(\sum_{i=1}^n |f_i(t)|^2 \right)^{p/2} dt \right)^{1/p}$$

$$= A_p \left\| \left(\sum_{i=1}^n |f_i|^2 \right)^{1/2} \right\|_p.$$

The upper estimate is similar. $\square$

We could paraphrase Proposition 10.2 by writing

$$\left\| \sum_{i=1}^n r_i(s) f_i(t) \right\|_{L_p([0,1]^2)} \approx \left\| \left(\sum_{i=1}^n |f_i(t)|^2 \right)^{1/2} \right\|_{L_p}.$$

The idea now is either to compare the left-hand side with, say, $\| \sum_{i=1}^n f_i \|_p$, as we might for an unconditionally convergent series, or to compare the right-hand side with $(\sum_{i=1}^n \| f_i \|_p^2)^{1/2}$ or with $(\sum_{i=1}^n \| f_i \|_p^p)^{1/p}$, as we might for disjointly supported sequences. As a first application of these ideas, we present a classical theorem due to Orlicz [111].

Theorem 10.3 (Orlicz's Theorem). *If $\sum_n f_n$ is unconditionally convergent in L_p, $1 \leq p < 2$, then $\sum_{n=1}^\infty \| f_n \|_p^2$ converges.*

Proof. Since $\sum f_n$ is unconditionally convergent, there is some constant K such that $\| \sum_{i=1}^n r_i(s) f_i \|_p \leq K$ for all n and all s. Thus, from Proposition 10.2, there is some constant C such that $\| (\sum_{i=1}^n |f_i|^2)^{1/2} \|_p \leq C$ for all n. All that remains is to compare this expression with $\sum_{i=1}^n \| f_n \|_p^2$.

But for $1 \leq p < 2$ we have

$$\left\| \left(\sum_{i=1}^{n} |f_i|^2 \right)^{1/2} \right\|_p^2 = \left(\int_0^1 \left(\sum_{i=1}^{n} |f_i(t)|^2 \right)^{p/2} dt \right)^{2/p}$$

$$\geq \sum_{i=1}^{n} \left(\int_0^1 |f_i(t)|^p \, dt \right)^{2/p} = \sum_{i=1}^{n} \|f_i\|_p^2,$$

where the inequality follows from the fact that $p/2 \leq 1$ and the triangle inequality in $L_{p/2}$ is reversed! $\square$

Orlicz's theorem reduces Banach's problem by eliminating one case.

Corollary 10.4. *If* $1 \leq p < 2 < r < \infty$, *or if* $1 \leq r < 2 < p < \infty$, *then there can be no isomorphism from* L_r *or* ℓ_r *into* L_p.

Proof. It's clearly enough to show that ℓ_r doesn't embed in L_p, and we've already settled this question in case $1 \leq r < 2 < p < \infty$.

Now suppose that $1 \leq p < 2 < r < \infty$ and that $T : \ell_r \to L_p$ is an isomorphism. Then, given $\sum a_n e_n$ in ℓ_r, the series $\sum a_n T e_n$ is unconditionally convergent in L_p. Hence, by Orlicz's theorem, we have

$$\infty > \sum |a_n|^2 \|T e_n\|_p^2 \geq \|T^{-1}\|^{-2} \sum |a_n|^2.$$

That is, $\sum |a_n|^2$ converges whenever $\sum |a_n|^r$ converges. This is clearly impossible for $r > 2$, as the series $\sum n^{-1/2}$ plainly demonstrates. $\square$

For $2 \leq p < \infty$, the conclusion of Orlicz's theorem changes.

Theorem 10.5. *If* $\sum_n f_n$ *is unconditionally convergent in* L_p, $2 \leq p < \infty$, *then* $\sum_{n=1}^{\infty} \|f_n\|_p^p$ *converges.*

Proof. For $2 \leq p < \infty$, we use a different trick:

$$\left\| \left(\sum_{i=1}^{n} |f_i|^2 \right)^{1/2} \right\|_p^p = \int_0^1 \left(\sum_{i=1}^{n} |f_i(t)|^2 \right)^{p/2} dt$$

$$\geq \int_0^1 \left(\sum_{i=1}^{n} |f_i(t)|^p \right)^{p/p} dt$$

$$= \sum_{i=1}^{n} \|f_i\|_p^p,$$

where here we've used the fact that $\| \cdot \|_{\ell_2} \geq \| \cdot \|_{\ell_p}$ for $p \geq 2$. $\square$

To this point, we have completely settled Banach's question in all but the cases $1 \le r < p < 2$ and $1 \le p < r < 2$. As we mentioned at the beginning of this chapter, the first case can be eliminated. The argument in this case is very similar to that used in the proof of Orlicz's theorem; this time we compute upper estimates for $\int_0^1 \left\| \sum_{i=1}^n r_i(s)f_i \right\|_p ds$.

Theorem 10.6. *For any* $f_1, \ldots, f_n \in L_p$,

$$\int_0^1 \left\| \sum_{i=1}^n r_i(s)f_i \right\|_p ds \le \left(\sum_{i=1}^n \|f_i\|_p^p \right)^{1/p} \qquad \text{for} \quad 1 \le p \le 2,$$

and

$$\int_0^1 \left\| \sum_{i=1}^n r_i(s)f_i \right\|_p ds \le B_p \left(\sum_{i=1}^n \|f_i\|_p^2 \right)^{1/2} \qquad \text{for} \quad 2 \le p < \infty.$$

Proof. As we've already seen,

$$\int_0^1 \left\| \sum_{i=1}^n r_i(s)f_i \right\|_p ds \le \left(\int_0^1 \left\| \sum_{i=1}^n r_i(s)f_i \right\|_p^p ds \right)^{1/p}$$

$$= \left(\int_0^1 \int_0^1 \left| \sum_{i=1}^n r_i(s)f_i(t) \right|^p ds\, dt \right)^{1/p}$$

$$\le B_p \left(\int_0^1 \left(\sum_{i=1}^n |f_i(t)|^2 \right)^{p/2} dt \right)^{1/p},$$

and $B_p = 1$ for $p \le 2$. All that remains is to estimate this last expression from above. But if $1 \le p \le 2$, then $p/2 \le 1$ and so

$$\int_0^1 \left(\sum_{i=1}^n |f_i(t)|^2 \right)^{p/2} dt \le \int_0^1 \sum_{i=1}^n |f_i(t)|^p dt = \sum_{i=1}^n \|f_i\|_p^p.$$

Whereas if $2 \le p < \infty$, then $p/2 \ge 1$ and so

$$\left(\int_0^1 \left(\sum_{i=1}^n |f_i(t)|^2 \right)^{p/2} dt \right)^{2/p} \le \left(\sum_{i=1}^n \int_0^1 |f_i(t)|^p dt \right)^{2/p} = \sum_{i=1}^n \|f_i\|_p^2$$

by the triangle inequality in $L_{p/2}$. $\square$

The choice of the expression $\int_0^1 \| \sum_{i=1}^n r_i(s)f_i \|_p ds$ as opposed to the expression $(\int_0^1 \| \sum_{i=1}^n r_i(s)f_i \|_p^p ds)^{1/p}$ in our last result is of little consequence.

We could, in fact, have used $(\int_0^1 \| \sum_{i=1}^n r_i(s)f_i \|_p^r ds)^{1/r}$ for any r. All such expressions are equivalent and, hence, all are equivalent to $\|(\sum_{i=1}^n |f_i|^2)^{1/2}\|_p$ (see [95, Theorem 1.e.13] or [147, III.A.18]).

Finally we're ready to deal with the last case that can be handled by elementary inequalities.

Corollary 10.7. *If* $1 \le r < p < 2$, *there can be no isomorphism from* L_r *or* ℓ_r *into* L_p.

Proof. As before, we only need to consider ℓ_r. So, suppose that $T : \ell_r \to L_p$ is an isomorphism. Then,

$$n^{1/r} = \int_0^1 \left\| \sum_{i=1}^n r_i(s)e_i \right\|_r ds \le \|T^{-1}\| \int_0^1 \left\| \sum_{i=1}^n r_i(s)T(e_i) \right\|_p ds$$

$$\le \|T^{-1}\| \left(\sum_{i=1}^n \|T(e_i)\|_p^p \right)^{1/p}$$

$$\le \|T^{-1}\| \|T\| n^{1/p}.$$

That is, $n^{1/r} \le Cn^{1/p}$ for all n, which is clearly impossible since $1/r > 1/p$. $\square$

Banach's approach to the case $1 \le r < p < 2$ was somewhat different from ours. Instead, he appealed to the *Banach–Saks theorem* [9]:

Theorem 10.8. *Every weakly null sequence* (f_n) *in* L_p, $1 < p \le 2$, *has a subsequence* (f_{n_i}) *with* $\| \sum_{i=1}^k f_{n_i} \|_p = O(k^{1/p})$.

Thus, if $1 < r < p < 2$, and if $T : \ell_r \to L_p$ is an isomorphism, then $(T(e_i))$ would have a subsequence satisfying $\| \sum_{i=1}^k T(e_{n_i}) \|_p = O(k^{1/p})$. But, just as above, the fact that T is an isomorphism implies that $\| \sum_{i=1}^k T(e_{n_i}) \|_p \approx \| \sum_{i=1}^k e_{n_i} \|_r = k^{1/r}$, which is a contradiction. This argument doesn't apply in the case $r = 1$ since (e_n) isn't weakly null in ℓ_1. But because ℓ_1 isn't reflexive, it can't possibly be isomorphic to a subspace of the reflexive space L_p for any $1 < p < \infty$.

The remaining case, $1 \le p < r < 2$, is substantially harder than the rest, but there's a payoff: For p and r in this range, L_p actually contains a closed subspace *isometric* to all of L_r. The proof requires several tools from probability theory that, taken one at a time, are not terribly difficult but, taken all at once, would require more time than we have.

Notes and Remarks

For more on unconditional convergence, see Day [29] or Diestel [33]. For more on unconditional bases, see also Lindenstrauss and Tzafriri [94, 95] and Megginson [100]. Proposition 10.2 and its relatives come to us partly through folklore but largely through the work of such giants as J. L. Krivine and B. Maurey. Such "square-function" inequalities are by now commonplace in harmonic analysis and probability theory as well as in Banach space theory. See also Zygmund [150]. For more on the unresolved case $1 \le p < r < 2$, as well as more on the work of Krivine, Maurey, and others, see Lindenstrauss and Tzafriri [95] and the *Memoir* by Johnson, et al. [81].

Exercises

1. If $\sum_n x_n$ converges unconditionally to x, prove that every rearrangement $\sum_n x_{\pi(n)}$ likewise converges to x.

2. If $\sum_n x_n$ is unconditionally convergent in X, show that the map $f : \{-1, 1\}^{\mathbb{N}} \to X$ defined by $f((\varepsilon_n)) = \sum_n \varepsilon_n x_n$ is continuous, where $\{-1, 1\}^{\mathbb{N}}$ is supplied with the metric $d((\varepsilon_n), (\rho_n)) = \sum_n |\varepsilon_n - \rho_n|/2^n$.

3. Suppose that the series $\sum_n a_n x_n$ converges in X for every bounded sequence of scalars (a_n). Use the Closed Graph theorem to prove that the map $T : \ell_\infty \to X$ defined by $T((a_n)) = \sum_n a_n x_n$ is continuous.

4. Let $1 < r < \infty$ and let $f(x, y)$ be a nonnegative function in $L_r([0, 1]^2)$. Prove that

$$\left[\int_0^1 \left(\int_0^1 f(x, y)\,dy \right)^r dx \right]^{1/r} \le \int_0^1 \left(\int_0^1 f(x, y)^r\,dx \right)^{1/r} dy.$$

5. Let $1 \le r \le p \le q < \infty$, and let $f_1, \ldots, f_n \in L_p$. Show that

$$\left(\sum_{i=1}^n \|f_i\|_p^q \right)^{1/q} \le \left\| \left(\sum_{i=1}^n |f_i|^q \right)^{1/q} \right\|_p$$

$$\le \left\| \left(\sum_{i=1}^n |f_i|^r \right)^{1/r} \right\|_p \le \left(\sum_{i=1}^n \|f_i\|_p^r \right)^{1/r}.$$

The inequality also holds for $q = \infty$ provided that we use $\max_{1 \le i \le n} \|f_i\|_p$ and $\left\| \max_{1 \le i \le n} |f_i| \right\|_p$ as the first two terms.

Chapter 11
Convexity

Several of the inequalities that we've generated in L_p spaces are generalizations of the parallelogram law. To see this, let's rewrite the usual parallelogram law for ℓ_2:

$$\|x + y\|_2^2 + \|x - y\|_2^2 + \| -x + y\|_2^2 + \| -x - y\|_2^2 = 4\left(\|x\|_2^2 + \|y\|_2^2\right).$$

That is, the average value of $\| \pm x \pm y\|_2^2$ over the four choices of signs is $\|x\|_2^2 + \|y\|_2^2$. In other words,

$$\int_0^1 \|r_1(t)\, x + r_2(t)\, y\|_2^2\, dt = \|x\|_2^2 + \|y\|_2^2.$$

Now the parallelogram law tells us something about the degree of "roundness" of the unit ball in ℓ_2. Indeed, if $\|x\|_2 = 1 = \|y\|_2$, and if $\|x - y\|_2 \geq \varepsilon > 0$, then

$$\|x + y\|_2^2 = 2\left(\|x\|_2^2 + \|y\|_2^2\right) - \|x - y\|_2^2 \leq 4 - \varepsilon^2.$$

Thus, $\|x + y\|_2 < 2$ for any two distinct points on the unit sphere in ℓ_2. That is, the midpoint $(x + y)/2$ has norm strictly less than 1 and so lies strictly inside the unit ball. In fact, we can even determine just how far inside the ball:

$$1 - \left\|\frac{x + y}{2}\right\|_2 \geq 1 - \sqrt{1 - \frac{\varepsilon^2}{4}} \geq \frac{\varepsilon^2}{8} = \delta.$$

It's not hard to see that every point on the chord joining x and y has norm strictly less than 1; thus, there can be no line segments on the sphere itself.

In this chapter we'll investigate various analogues of the parallelogram law and their consequences in a general normed linear space X.

107

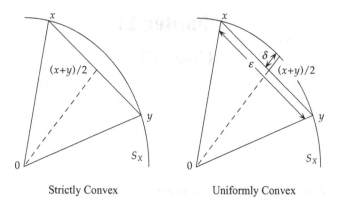

Strictly Convex   Uniformly Convex

Strict Convexity

It's often convenient to know whether the triangle inequality is strict for non-collinear points in a given normed space. We say that a normed space X is *strictly convex* if

$$\|x + y\| < \|x\| + \|y\|$$

whenever x and y are not parallel. (That is, when they are not on the same line through 0 and hence not multiples of one another. The triangle inequality is always strict if y is a negative multiple of x, whereas it's always an equality if y is a positive multiple of x.) Since strict convexity is more accurately attributed to the norm in X, we sometimes say instead that X has a *strictly convex norm*. Like the parallelogram law itself, this is very much an isometric property, as we'll see shortly.

Let's begin with several easy examples.

(a) A review of the proof of the triangle inequality for Hilbert space (and the case for equality in the Cauchy–Schwarz inequality) shows that any Hilbert space is strictly convex. Also, for any $1 < p < \infty$, it follows from the case for equality in Minkowski's inequality that $L_p(\mu)$ is strictly convex. If you're still skeptical, we'll give several different proofs of these facts before we're done.

(b) Consider ℓ_∞^2; that is, $\mathbb{R}^2$ under the norm $\|(x, y)\|_\infty = \max\{|x|, |y|\}$. It's easy to see that ℓ_∞^2 is not strictly convex. For example, the vectors $(1, 1)$ and $(1, -1)$ are not parallel and yet $\|(1, 1) \pm (1, -1)\|_\infty = 2 = \|(1, 1)\|_\infty + \|(1, -1)\|_\infty$. What's more, the entire line segment joining $(1, 1)$ and $(1, -1)$ lies on the unit sphere since, for any $0 \le \lambda \le 1$, we have $\|\lambda(1, 1) + (1 - \lambda)(1, -1)\|_\infty = \|(1, 2\lambda - 1)\|_\infty = 1$.

For later reference, let's rephrase this observation in two different ways. First, the point $(2, 0)$ is not in the unit ball of ℓ_∞^2, a compact convex subset of ℓ_∞^2. It follows that $(2, 0)$ must have a nearest point on the sphere. In fact, it has many; every point on the segment joining $(1, 1)$ and $(1, -1)$ is nearest $(2, 0)$: $\|(2, 0) - (1, y)\|_\infty = \|(1, -y)\|_\infty = 1$ for $-1 \leq y \leq 1$.

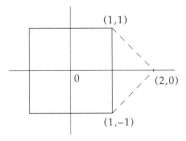

As a second restatement, consider the element $(1, 0) \in \ell_1^2$, the predual of ℓ_∞^2. Notice that every functional of the form $(1, y) \in \ell_\infty^2, -1 \leq y \leq 1$, norms $(1, 0)$: $\langle (1, 0), (1, y) \rangle = 1 = \|(1, 0)\|_1 \|(1, y)\|_\infty$.

It follows that any space that contains an isometric copy of ℓ_∞^2 is likewise not strictly convex. Thus, c_0, ℓ_∞, L_∞, and $C[0, 1]$ are not strictly convex. In L_∞, for example, the functions $\chi_{[0,1/2]}$ and $\chi_{[1/2,1]}$ span an isometric copy of ℓ_∞^2.

(c) It's nearly obvious that ℓ_1 and L_1 are not strictly convex since the triangle inequality is an equality on any pair of vectors of the same sign (coordinatewise or pointwise). Again, for later reference, let's state this fact in a couple of different ways. In L_1, consider the "positive face" of the unit sphere:

$$K = \left\{ f \in L_1 : f \geq 0, \int_0^1 f = 1 \right\}.$$

Clearly, K is a closed *convex* subset of the unit sphere of L_1 (it's the intersection of two closed convex sets) and K contains *lots* of line segments! (Why?) Also note that *every* point in K is distance 1 away from $0 \notin K$. Finally, notice that the functional $f \mapsto \int_0^1 f$ attains its norm at *every* point of K.

As a second example, consider

$$K = \left\{ x \in \ell_1 : \sum_{n=1}^\infty \left(1 - \frac{1}{n+1}\right) x_n = 1 \right\}.$$

Once more, K is a closed convex set in ℓ_1 (it's a hyperplane) and $0 \notin K$. This time, however, there is no point in K nearest 0; that is, no element of smallest norm in K. Indeed, it's easy to check that $\text{dist}(0, K) = \|(1 - \frac{1}{n+1})_{n=1}^\infty\|_\infty^{-1} = 1$, while for every $x \in K$ we have

$$1 = \left|\sum_{n=1}^\infty \left(1 - \frac{1}{n+1}\right)x_n\right| < \sum_{n=1}^\infty |x_n| = \|x\|_1.$$

Said still another way, this same calculation shows that the functional $f = (1 - \frac{1}{n+1})_{n=1}^\infty \in \ell_\infty$ doesn't attain its norm on the sphere of ℓ_1.

(d) Strict convexity is very much an isometric property and isn't typically preserved by isomorphisms or equivalent renormings. For example, consider the norm

$$\||(x, y)\|| = \max\{2|x|, \|(x, y)\|_2\}$$

on $\mathbb{R}^2$. Clearly, $\|| \cdot \||$ is equivalent to the Euclidean norm $\| \cdot \|_2$. But $\|| \cdot \||$ isn't strictly convex, for if we take $(1, 0), (1, 1) \in \mathbb{R}^2$, then $\||(1, 0)\|| = 2$, $\||(1, 1)\|| = 2$, while $\||(1, 0) + (1, 1)\|| = \||(2, 1)\|| = 4$.

It's about time we gave a formal proof or two. First, let's give two equivalent characterizations of strict convexity.

Theorem 11.1. *X is strictly convex if and only if either of the following holds:*

(i) *For $x \neq y$ in X with $\|x\| = 1 = \|y\|$ we have $\|\frac{x+y}{2}\| < 1$.*
(ii) *For $1 < p < \infty$ and $x \neq y$ in X we have $\|\frac{x+y}{2}\|^p < \frac{\|x\|^p + \|y\|^p}{2}$.*

Proof. Condition (i) is surely implied by strict convexity; if $x \neq y$ are norm one vectors, then either x and y are nonparallel, or else $y = -x$, and in either case (i) follows. Suppose, on the other hand, that (i) holds but that we can find nonparallel vectors x and y in X with $\|x + y\| = \|x\| + \|y\|$. We may clearly assume that $0 < \|x\| \leq \|y\|$. Thus,

$$\left\|\frac{x}{\|x\|} + \frac{y}{\|y\|}\right\| \geq \left\|\frac{x}{\|x\|} + \frac{y}{\|x\|}\right\| - \left\|\frac{y}{\|x\|} - \frac{y}{\|y\|}\right\|$$

$$= \frac{\|x\| + \|y\|}{\|x\|} - \|y\|\left(\frac{1}{\|x\|} - \frac{1}{\|y\|}\right) = 2,$$

which contradicts (i).

Next, suppose that X is strictly convex and let $1 < p < \infty$. Given non-parallel vectors x and y in X, we have

$$\left\| \frac{x+y}{2} \right\|^p < \left(\frac{\|x\| + \|y\|}{2} \right)^p \leq \frac{\|x\|^p + \|y\|^p}{2}$$

since the function $|t|^p$ is convex. Similarly, if $y = tx$, $t \neq 1$, then

$$\left\| \frac{x+y}{2} \right\|^p = \left| \frac{1+t}{2} \right|^p \|x\|^p < \frac{1 + |t|^p}{2} \|x\|^p = \frac{\|x\|^p + \|y\|^p}{2}$$

because the function $|t|^p$ is *strictly* convex. Thus, (ii) holds. That (ii) implies strict convexity now follows easily from (i). □

That L_p is strictly convex for $1 < p < \infty$ follows from (ii) and the fact that $|t|^p$ is strictly convex. We'll give another proof of this later.

It's clear from our pictures that if X is strictly convex, then S_X, the unit sphere in X, contains no nontrivial line segments. What this means is that each point of S_X "sticks out" from its neighbors. A fancier way to describe this is to say that each point of S_X is an *extreme point* of B_X, the closed unit ball of X. A point x in a convex set K is said to be an extreme point if x cannot be written as a nontrivial convex combination of distinct points in K. Thus, x is an extreme point of K if and only if

$$x = (y + z)/2, \, y, \, z \in K \Longrightarrow y = z = x.$$

Now a convex set need not have extreme points, as the closed unit ball of c_0 will attest, but if it does they have to live on the boundary of the set. In particular, any extreme point of B_X must live in S_X. Indeed, if $\|x\| < 1$, then $\|x\| < (1 + \delta)^{-1}$ for some $0 < \delta < 1$. Thus, $\|(1 \pm \delta)x\| < 1$ and we would then have $x = \frac{1}{2}[(1 + \delta)x + (1 - \delta)x]$. Hence, x is not extreme. Now if X is strictly convex and if $\|x\| = 1$, then x must be an extreme point for B_X by condition (i) of our last result. In a strictly convex space, then, not only does B_X have extreme points, but every point in S_X is extreme.

But we can say even more: If X is strictly convex, then each point of S_X is even an *exposed point* of B_X. That is, for each $x \in S_X$, there exists a norm one functional $f_x \in X^*$ that "exposes" x; that is, $f_x(x) = 1$ while $f_x(y) < 1$ for all other $y \in B_X$. Now an exposed point is also an extreme point, for if $x = (y + z)/2$, then $1 = f_x(x) = (f_x(y) + f_x(z))/2$, and hence at least one of $f_x(y) \geq 1$ or $f_x(z) \geq 1$ holds.

Theorem 11.2. *X is strictly convex if and only if any one of the following holds:*

(iii) *S_X contains no line segment.*
(iv) *Every point of S_X is an extreme point of B_X.*
(v) *Every point of S_X is an exposed point of B_X.*

Proof. It's nearly obvious that strict convexity is equivalent to each of (iii) and (iv), and so let's concentrate on the equivalence of strict convexity with (v). As we've already pointed out, (v) implies (iv); thus, we just need to show that something on the list implies (v). Let's go for the obvious: Given $x \in S_X$, the Hahn–Banach theorem supplies a norm one functional $f \in X^*$ with $f(x) = 1$. What happens if $f(y) = 1$ for some $y \neq x$ in B_X? Well, for one, we'd have to have $\|y\| = 1$ since $1 = f(y) \leq \|y\|$. But then, $f((x + y)/2) = 1$ likewise forces $\|(x + y)/2\| = 1$ in violation of the strict convexity of X. Consequently, we must have $f(y) < 1$ for every $y \neq x$ in B_X. $\square$

Nearest Points

We next turn our attention, all too briefly, to the issue of nearest points. Given a nonempty set K in a normed linear space X and a point $x_0 \in X$, we say that the point $x \in K$ is nearest x_0 if $\|x - x_0\| = \inf_{y \in K} \|y - x_0\|$. Obviously, if $x_0 \in K$, then x_0 is its own (unique) nearest point in K. If K is closed, then this infimum exists and is positive for any $x_0 \notin K$, but it need not be attained, in general. Moreover, even if nearest points exist, they need not be unique, as our earlier examples point out. The questions that arise, then, are; Under what conditions on K will nearest points always exist? When are they unique? If each x_0 has a unique nearest point $p(x_0) \in K$, what can we say about the (typically nonlinear) "nearest point map" p?

Now it's an easy exercise to show that if K is *compact*, then each point $x_0 \notin K$ has a nearest point in K. In fact, it's not much harder to show that the same holds for a *weakly compact* set K. Once we bring the weak topology into the picture, though, it's typical to add the requirement that K be convex since weak- and norm-closures coincide for convex sets. (This is a consequence of the Hahn–Banach theorem.) For these reasons, nearest-point problems are often stated in terms of closed convex sets. We'll adopt this convention wholeheartedly. For now, let's settle for a few simple observations.

Theorem 11.3. *Let K be a nonempty, compact, convex subset of a normed linear space X.*

(a) *Each $x_0 \in X$ has a nearest point in K.*

(b) *If X is strictly convex, then there is a unique point $p(x_0) \in K$ nearest to x_0. Moreover, the map $x_0 \mapsto p(x_0)$ is continuous.*

Proof. A carefully constructed proof will give all three conclusions at once. Here goes: Let $x_0 \notin K$, and let $d = \inf_{y \in K} \|y - x_0\| > 0$. For each n, consider

$$K_n = \left\{ x \in K : d \le \|x - x_0\| \le d + \tfrac{1}{n} \right\} = K \cap \left(x_0 + \left(d + \tfrac{1}{n} \right) B_X \right).$$

Each K_n is a nonempty, closed, convex subset of K, and $K_{n+1} \subset K_n$ for all n. Thus, $\bigcap_{n=1}^{\infty} K_n \neq \emptyset$. Clearly, any point $y \in \bigcap_{n=1}^{\infty} K_n$ is a nearest point to x_0 in K.

Now if x, $y \in K$ are each nearest to x_0, then so is their midpoint $z = (x + y)/2$ since

$$d \le \|z - x_0\| = \left\| \tfrac{1}{2}(x - x_0) + \tfrac{1}{2}(y - x_0) \right\| \le d.$$

Thus, if X is strictly convex, then we must have $x - x_0 = y - x_0 = z - x_0$; hence, $x = y = z$. That is, there can be at most one nearest point. What's more, this tells us that *the diameters of the sets K_n must tend to 0* when X is strictly convex, for otherwise we could find two points nearest to x_0.

Finally, suppose that X is strictly convex and let (x_n) be a sequence in X converging to x_0. There's no great harm in supposing that $\|x_n - x_0\| \le 1/2n$ (this will simplify the proof). It then follows that $d_n = \inf_{y \in K} \|y - x_n\| \le d + 1/2n$ and hence that $\|x_0 - p(x_n)\| \le \|x_n - p(x_n)\| + \|x_n - x_0\| \le d_n + 1/2n \le d + 1/n$, where $p(x_n)$ is the unique point in K nearest x_n. Consequently, we must have $p(x_n) \in K_n$. Because the diameter of K_n tends to 0, we get $p(x_n) \to p(x_0)$. $\square$

Smoothness

Related to the notion of strict convexity is the notion of smoothness. We say that a normed space Y is *smooth* if, for each $0 \neq y \in Y$, there exists a *unique* norm one functional $f \in Y^*$ such that $f(y) = \|y\|$. Of course, the Hahn–Banach theorem ensures the existence of at least one such functional f; what's at issue here is uniqueness. In any case, it's clearly enough to check only norm one vectors y when testing for smoothness. As it happens, a normed space is smooth if and only if its norm has directional derivatives in each direction, which accounts for the name of the property. We won't pursue this further; our interest in smoothness is as a dual property to strict convexity.

It's immediate that a Hilbert space H is smooth. Indeed, each vector norms itself in the sense that $\langle x, x \rangle = \|x\|^2$ for all $x \in H$, and uniqueness follows from (the converse to) the Cauchy–Schwarz inequality. Similarly, it follows from the converse to Hölder's inequality that L_p is smooth for $1 < p < \infty$. In this case, each $f \in L_p$ is normed by the function $|f|^{p-1}\mathrm{sgn}\, f \in L_q$.

Our examples at the beginning of this chapter should convince you that c_0, ℓ_1, ℓ_∞, L_1, L_∞, and $C[0, 1]$ all fail to be smooth. In c_0, for instance, the vector $e_1 + e_2$ is normed by e_1, e_2, and $\frac{1}{2}(e_1 + e_2) \in \ell_1$. Similarly, the vector $e_1 \in \ell_1$ is normed by e_1, $e_1 + e_2$, and $e_1 + e_2 + e_3 \in \ell_\infty$.

Theorem 11.4. *If X^* is strictly convex, then X is smooth. If X^* is smooth, then X is strictly convex.*

Proof. If X is not smooth, then we can find a norm one vector $x \in X$ that is normed by two distinct norm one functionals $f \neq g \in X^*$. But then, $\|f + g\| \leq 2$ and $(f + g)(x) = 2$; hence, $\|f + g\| = 2$. This denies the strict convexity of X^*.

If X is not strictly convex, then we can find a norm one vector $x \in X$ that is not an exposed point. Thus, if we choose a norm one $f \in X^*$ with $f(x) = 1$, then we must have $f(y) = 1$ for some other norm one $y \neq x$. But then, $\hat{x}$, $\hat{y} \in X^{**}$ both norm $f \in X^*$, denying the smoothness of X^*. $\square$

Smoothness and strict convexity aren't quite dual properties: There are examples of strictly convex spaces whose duals fail to be smooth. But there is at least one easy case where full duality has to hold:

Corollary 11.5. *If X is reflexive, then X is strictly convex (resp., smooth) if and only if X^* is smooth (resp., strictly convex).*

Notice that the spaces L_p, $1 < p < \infty$, and any Hilbert space H necessarily enjoy *both* properties.

Uniform Convexity

A normed linear space X is said to be *uniformly convex* if, for each $\varepsilon > 0$, there is a $\delta = \delta(\varepsilon) > 0$ such that

$$\|x\| \leq 1,\ \|y\| \leq 1,\ \|x - y\| \geq \varepsilon \implies \left\|\frac{x + y}{2}\right\| \leq 1 - \delta. \quad (11.1)$$

Obviously, any uniformly convex space is also strictly convex, but there are strictly convex spaces that aren't uniformly convex. An easy compactness argument will convince you that if X is finite dimensional, then X is strictly convex if and only if X is uniformly convex. Just as with strict convexity, uniform convexity is actually a property of the norm on X, and so it might be more appropriate to say that X has a *uniformly convex norm* whenever (11.1) holds.

As in the strictly convex case, we could reformulate our definition in terms of pairs of vectors x, y with $\|x\| = \|y\| = 1$ and $\|x - y\| = \varepsilon$. Likewise, we could use some power of the norm, say $\| \cdot \|^p$ for $p > 1$. The details in this case are rather tedious and not particularly important to our discussion, so we will omit them.

As we saw at the beginning of this chapter, the parallelogram law implies that every Hilbert space is uniformly convex. In fact, we even computed $\delta(\varepsilon) = 1 - (1 - \varepsilon^2/4)^{1/2}$. We will see that uniformly convex spaces share many of the geometric niceties of Hilbert space. Our primary aim is to prove the famous result, due to Clarkson in 1936 [25], that L_p is uniformly convex whenever $1 < p < \infty$. Using this observation, we will give a geometric proof (essentially devoid of measure theory) that $L_p^* = L_q$. In fact, we will show that every uniformly convex Banach space is necessarily reflexive.

Note that because L_1 and L_∞ fail to be even strictly convex, they can't be uniformly convex. The same is true of c_0, ℓ_1, ℓ_∞, and $C[0, 1]$. Also note that uniform convexity is very much an isometric property; the equivalent renorming of $\mathbb{R}^2$ we gave earlier that fails to be strictly convex obviously also fails to be uniformly convex. Along similar lines, it's easy to check that the expression $\||x\|| = \|x\|_1 + \|x\|_2$ defines an equivalent strictly convex norm on ℓ_1 but that ℓ_1 supplied with this norm can't be uniformly convex since it's not reflexive.

We begin with a simple but very useful observation.

Lemma 11.6. *X is uniformly convex if and only if, for all pairs of sequences (x_n), (y_n) with $\|x_n\| \leq 1$, $\|y_n\| \leq 1$, we have*

$$\left\| \frac{x_n + y_n}{2} \right\| \to 1 \implies \|x_n - y_n\| \to 0.$$

Proof. One the one hand, if X is uniformly convex, and if $\|(x_n + y_n)/2\| \to 1$, then we must have $\|x_n - y_n\| \to 0$, for otherwise, $\|x_n - y_n\| \geq \varepsilon$ implies $\|(x_n + y_n)/2\| \leq 1 - \delta$.

On the other hand, if X is not uniformly convex, then we can find an $\varepsilon > 0$ and sequences (x_n) and (y_n) in B_X such that $\|x_n - y_n\| \geq \varepsilon$ whereas $\|(x_n + y_n)/2\| \to 1$. $\square$

Corollary 11.7. *Let X be uniformly convex. If (x_n) in X satisfies $\|x_n\| \leq 1$ and $\|(x_n + x_m)/2\| \to 1$ as $m, n \to \infty$, then (x_n) is Cauchy!*

Note that the condition $\|(x_n + x_m)/2\| \to 1$ also forces $\|x_n\| \to 1$ by the triangle inequality: $\|(x_n + x_m)/2\| \leq (\|x_n\| + \|x_m\|)/2 \leq 1$. Thus, if X is complete, (x_n) converges to some norm one element of X.

Our last two results can be used to prove several classical convergence theorems, but we'll settle for just one such result: In the case $X = L_p$, it's sometimes called the Radon–Riesz theorem.

Theorem 11.8. *Let X be a uniformly convex Banach space, and suppose that (x_n) in X satisfies $x_n \xrightarrow{w} x$ and $\|x_n\| \to \|x\|$. Then $\|x_n - x\| \to 0$.*

Proof. If $\|x\| = 0$ there's nothing to show, and so we may suppose that $x \neq 0$. Now, since $\|x_n\| \to \|x\|$, it's easy to see that we can normalize our sequence: If we set $y = x/\|x\|$ and $y_n = x_n/\|x_n\|$, then $x_n \xrightarrow{w} x$ implies that $y_n \xrightarrow{w} y$; also, $\|y_n - y\| \to 0$ will imply that $\|x_n - x\| \to 0$.

Next, choose a norm one functional $f \in X^*$ with $f(y) = 1$. Then,

$$f\left(\frac{y_n + y_m}{2}\right) \leq \left\|\frac{y_n + y_m}{2}\right\| \leq 1.$$

But $f(y_n) \to f(y) = 1$ as $n \to \infty$, and so we must have $\|(y_n + y_m)/2\| \to 1$ as $m, n \to \infty$. From Corollary 11.7, this means that (y_n) is Cauchy and, hence, (y_n) converges in norm to some point in X. But, because norm convergence implies weak convergence, and since weak limits are unique, we must have $y_n \to y$ in norm. $\square$

Using a very similar argument, we can give a short proof of an interesting fact, due to Milman in 1938 [103] and Pettis in 1939 [118]. (If you're not familiar with *nets*, just pretend they're sequences; you won't be far from the truth.)

Theorem 11.9. *A uniformly convex Banach space is reflexive.*

Proof. Let X be uniformly convex and let $x^{**} \in X^{**}$ with $\|x^{**}\| = 1$. We need to show that $x^{**} = \hat{x}$ for some $x \in X$. Now, since B_X is weak* dense in $B_{X^{**}}$, we can find a net (x_α) in X with $\|x_\alpha\| \leq 1$ such that $x_\alpha \xrightarrow{w^*} x^{**}$. But since $\|x_\alpha\| \leq 1 = \|x^{**}\|$, it follows that $\|x_\alpha\| \to \|x^{**}\|$. A slight variation on Theorem 11.8 shows that (x_α) is Cauchy in X and hence must converge to some $x \in X$. It follows that $x^{**} = \hat{x}$. $\square$

Clarkson's Inequalities

To round out our discussion of uniform convexity, we next prove *Clarkson's theorem*:

Theorem 11.10. L_p *is uniformly convex for* $1 < p < \infty$.

As it happens, the proof of Clarkson's theorem is quite easy for $2 \leq p < \infty$ and not quite so easy for $1 < p < 2$. For this reason, dozens of proofs have been given. We have the luxury of selecting bits and pieces from several of them. Now Clarkson proved several inequalities for the L_p norm that mimic the parallelogram law. We will do the same but without using his original proofs. It should be pointed out that all of the results we'll give in this section hold for every $L_p(\mu)$ space (after all, this has more to do with the L_p norm than with measure theory).

We first prove Clarkson's theorem for the case $2 \leq p < \infty$; this particular proof is due to Boas in 1940 [17].

Lemma 11.11. *Given* $2 \leq p < \infty$ *and real numbers* a *and* b, *we have*

$$|a + b|^p + |a - b|^p \leq 2^{p-1}(|a|^p + |b|^p).$$

Proof. We've seen this inequality before (more or less). See if you can fill in the reasons behind the following arithmetic:

$$\begin{aligned}
(|a + b|^p + |a - b|^p)^{1/p} &\leq (|a + b|^2 + |a - b|^2)^{1/2} \\
&= 2^{1/2}(|a|^2 + |b|^2)^{1/2} \\
&\leq 2^{1/2}2^{1/2-1/p}(|a|^p + |b|^p)^{1/p} \\
&= 2^{1-1/p}(|a|^p + |b|^p)^{1/p}. \quad \square
\end{aligned}$$

Theorem 11.12. *For* $2 \leq p < \infty$ *and any* $f, g \in L_p$, *we have*

$$\|f + g\|_p^p + \|f - g\|_p^p \leq 2^{p-1}\left(\|f\|_p^p + \|g\|_p^p\right). \tag{11.2}$$

Corollary 11.13. L_p *is uniformly convex for* $2 \leq p < \infty$.

Proof. If $f, g \in L_p$ with $\|f\|_p \leq 1$, $\|g\|_p \leq 1$, and $\|f - g\|_p \geq \varepsilon$, then

$$\|f + g\|_p^p \leq 2^{p-1} \cdot 2 - \varepsilon^p = 2^p \left(1 - \left(\frac{\varepsilon}{2}\right)^p\right).$$

That is, $\delta(\varepsilon) = 1 - (1 - (\frac{\varepsilon}{2})^p)^{1/p}$, and this is known to be exact. $\quad \square$

Unfortunately, inequality (11.2) *reverses* for $1 < p \leq 2$, and so we need a different proof in this case; the one we'll give is due to Friedrichs from 1970 [50]. As with Boas's proof, we begin with a pointwise inequality.

Lemma 11.14. *If* $1 < p \leq 2$, $q = p/(p-1)$, *and* $0 \leq x \leq 1$, *then*

$$(1+x)^q + (1-x)^q \leq 2(1+x^p)^{q-1}. \tag{11.3}$$

Proof. Consider the function

$$f(\alpha, x) = (1 + \alpha^{1-q}x)(1 + \alpha x)^{q-1} + (1 - \alpha^{1-q}x)(1 - \alpha x)^{q-1}$$

for $0 \leq \alpha \leq 1$ and $0 \leq x \leq 1$. Then, $f(1, x)$ is the left-hand side of (11.3) and $f(x^{p-1}, x)$ is the right-hand side of (11.3) because $(p-1)(q-1) = 1$. Thus, since $1 \geq x^{p-1}$, we want to show that $\partial f/\partial \alpha \leq 0$. But

$$\frac{\partial f}{\partial \alpha} = (q-1)x \cdot (1 - \alpha^{-q}) \cdot \left[(1 + \alpha x)^{q-2} - (1 - \alpha x)^{q-2}\right],$$

and $(1 - \alpha^{-q}) \leq 0$ because $\alpha \leq 1$, and $(1 + \alpha x)^{q-2} - (1 - \alpha x)^{q-2} \geq 0$ because $q \geq 2$. $\square$

For $p \geq 2$, the inequality in (11.3) holds with the roles of p and q exchanged. Our proof shows that the inequality in (11.3), as written, reverses for $p \geq 2$.

For $1 < p \leq 2$, the inequality in (11.3) easily implies that

$$|a+b|^q + |a-b|^q \leq 2(|a|^p + |b|^p)^{q-1}$$

for all real numbers a and b. But this time we can't simply integrate both sides to arrive at an L_p result. Instead, we'll need Friedrichs' clever extension of this inequality to L_p.

Theorem 11.15. *For* $1 < p \leq 2$, $q = p/(p-1)$, *and any* $f, g \in L_p$, *we have*

$$\|f+g\|_p^q + \|f-g\|_p^q \leq 2\left(\|f\|_p^p + \|g\|_p^p\right)^{q-1}.$$

Proof. First notice that

$$\|f\|_p^q = \left(\int |f|^p\right)^{1/(p-1)} = \left(\int |f|^{q(p-1)}\right)^{1/(p-1)} = \||f|^q\|_{p-1}$$

and $0 < p - 1 < 1$! Now, since the triangle inequality reverses in L_{p-1},

$$
\begin{aligned}
\|f + g\|_p^q + \|f - g\|_p^q &= \||f + g|^q\|_{p-1} + \||f - g|^q\|_{p-1} \\
&\leq \||f + g|^q + |f - g|^q\|_{p-1} \\
&\leq 2\|\left(|f|^p + |g|^p\right)^{q-1}\|_{p-1} \\
&= 2\left(\|f\|_p^p + \|g\|_p^p\right)^{q-1},
\end{aligned}
$$

where the last equality follows from the fact that $(q - 1)(p - 1) = 1$. □

Corollary 11.16. *L_p is uniformly convex for $1 < p \leq 2$.*

Notice that if $f, g \in L_p$, $1 < p \leq 2$, with $\|f\|_p \leq 1$, $\|g\|_p \leq 1$, and $\|f - g\|_p \geq \varepsilon$, then, just as before, we'd get $\delta(\varepsilon) = 1 - (1 - (\frac{\varepsilon}{2})^q)^{1/q} \approx \frac{1}{q}(\frac{\varepsilon}{2})^q$, where $q = p/(p - 1) \geq 2$. This, however, is too small. And ε^p is too big. It's known that $\delta(\varepsilon) \approx (p - 1)\varepsilon^2/8$ is (asymptotically) the correct value for L_p, $1 < p \leq 2$. (Notice that this is essentially the value of $\delta(\varepsilon)$ for Hilbert space; the point here is that no space can be "more convex" than Hilbert space.)

For the sake of completeness, we list all of Clarkson's inequalities (well, half of them anyway). Let $2 \leq p < \infty$ and $q = p/(p - 1)$. Then, for any f, $g \in L_p$, we have

$$2\left(\|f\|_p^p + \|g\|_p^p\right) \leq \|f + g\|_p^p + \|f - g\|_p^p \leq 2^{p-1}\left(\|f\|_p^p + \|g\|_p^p\right),$$

$$\|f + g\|_p^p + \|f - g\|_p^p \leq 2\left(\|f\|_p^q + \|g\|_p^q\right)^{p-1},$$

$$2\left(\|f\|_p^p + \|g\|_p^p\right)^{q-1} \leq \|f + g\|_p^q + \|f - g\|_p^q.$$

These inequalities all *reverse* for $1 < p \leq 2$. In particular, all reduce to the parallelogram identity when $p = 2$.

An Elementary Proof That $L_p^* = L_q$

In our discussion of the duality between strict convexity and smoothness, we made the claim that smoothness had something to do with differentiability of the norm. It should come as no surprise, then, that uniform convexity is dual to an even stronger differentiability property of the norm. Rather than formalize the condition, we'll settle for a few simple observations. Most of these results are due to McShane from 1950 [96].

We begin with a general principle: If a linear functional attains its norm at a point of differentiability of the norm in X, then its value is actually given by an appropriate derivative. This is the content of

Lemma 11.17 (McShane's Lemma). *Let X be a normed linear space and let $T \in X^*$. Suppose that $f, g \in X$ satisfy*

(i) $\|g\| = 1$ *and* $T(g) = \|T\|$,

(ii) $\lim_{t \to 0} \frac{\|g + tf\|^p - \|g\|^p}{pt}$ *exists for some $p \geq 1$.*

Then $T(f) = \|T\| \cdot \lim_{t \to 0} \frac{\|g + tf\|^p - \|g\|^p}{pt}$.

It should be pointed out that the pth power in (ii) is purely cosmetic–the case $p = 1$ is enough–but it will make life easier when we apply the lemma to the L_p norm. The number $\lim_{t \to 0}(\|g + tf\|^p - \|g\|^p)/pt$ is the directional derivative of $\| \cdot \|_p^p$ at g in the direction of f.

Proof. First we use l'Hôpital's rule to compute $T(f)$ as a derivative.

$$\lim_{t \to 0} \frac{(T(g + tf))^p - (T(g))^p}{pt} = \lim_{t \to 0} \frac{(T(g) + t\,T(f))^p - (T(g))^p}{pt}$$
$$= \lim_{t \to 0} (T(g) + t\,T(f))^{p-1} \cdot T(f)$$
$$= (T(g))^{p-1} \cdot T(f)$$
$$= \|T\|^{p-1} \cdot T(f).$$

Now, since $\|T\|\|g\| = T(g)$ and $\|T\|\|g + tf\| \geq T(g + tf)$, we get

$$\lim_{t \to 0^+} \|T\|^p \cdot \frac{\|g + tf\|^p - \|g\|^p}{pt} \geq \lim_{t \to 0^+} \frac{(T(g + tf))^p - (T(g))^p}{pt}$$
$$= \|T\|^{p-1} \cdot T(f)$$
$$= \lim_{t \to 0^-} \frac{(T(g + tf))^p - (T(g))^p}{pt}$$
$$\geq \lim_{t \to 0^-} \|T\|^p \cdot \frac{\|g + tf\|^p - \|g\|^p}{pt}.$$

$\square$

McShane's lemma gives us a schedule: First we show that each $T \in L_p^*$, $1 < p < \infty$, actually attains its norm, and then we compute the limit given in (ii). Curiously, that each element of L_p^* attains its norm is equivalent to the fact that L_p is reflexive – which doesn't actually require knowing anything about the dual space L_p^*!

First, a general fact:

Lemma 11.18. *If X is a uniformly convex Banach space, then each $T \in X^*$ attains its norm at a unique $g \in X$, $\|g\| = 1$.*

Proof. Choose (g_n) in X, $\|g_n\| = 1$ such that $T(g_n) \to \|T\|$. Then

$$2 \geq \|g_n + g_m\| \geq \|T\|^{-1}|T(g_n + g_m)| \to 2$$

as $m, n \to \infty$. Thus, (g_n) is Cauchy and hence converges to some $g \in X$. Clearly, $T(g) = \|T\|$ and $\|g\| = 1$. Uniform (or even strict) convexity tells us that g must be unique. □

Next, let's do the calculus step:

Lemma 11.19. *Let $f, g \in L_p$, $1 < p < \infty$. Then*

$$\lim_{t \to 0} \frac{\|g + tf\|^p - \|g\|^p}{pt} = \int f \cdot |g|^{p-1} \operatorname{sgn} g.$$

Proof. For any $a, b \in \mathbb{R}$, the function $\varphi(t) = |a + bt|^p$ is convex and has $\varphi'(t) = p|a + bt|^{p-1} \cdot b \cdot \operatorname{sgn}(a + bt)$. The limit now follows from the Dominated (or even Monotone) Convergence theorem:

$$\lim_{t \to 0} \frac{\|g + tf\|^p - \|g\|^p}{pt} = \int \lim_{t \to 0} \frac{|g(x) + tf(x)|^p - |g(x)|^p}{pt} dx$$

$$= \int f \cdot |g|^{p-1} \operatorname{sgn} g.$$

Alternatively, we could use the standard approach: Let

$$F(t) = \|g + tf\|_p^p = \int |g(x) + tf(x)|^p dx = \int \psi(x, t) dx.$$

Then, F is differentiable and $F'(t) = \int \psi_t(x, t) dx$ provided that ψ_t exists and is integrable. But, for $-1 \leq t \leq 1$, for instance, we have

$$|\psi_t(x, t)| = \left| p f(x)|g(x) + tf(x)|^{p-1} \operatorname{sgn}(g(x) + tf(x)) \right|$$

$$\leq p \, 2^{p-1} \left(|f(x)||g(x)|^{p-1} + |f(x)|^p \right),$$

which is integrable because $|g|^{p-1} \in L_q$. □

Now, given $T \in L_p^*$, take $g \in L_p$ with $\|g\|_p = 1$ and $T(g) = \|T\|$. Then, for every $f \in L_p$,

$$T(f) = \lim_{t \to 0} \|T\| \frac{\|g + tf\|^p - \|g\|^p}{pt}$$

$$= \|T\| \int f \cdot |g|^{p-1} \operatorname{sgn} g.$$

That is, T is represented by integration against $\|T\| |g|^{p-1} \operatorname{sgn} g \in L_q$, and, of course, $\| |g|^{p-1} \|_q = \|g\|_p^{p-1} = 1$. This proves that $L_p^* = L_q$.

Notes and Remarks

Our presentation in this chapter borrows from numerous sources; such as the books by Beauzamy [11], Day [29], Diestel [31, 33], Hewitt and Stromberg [69], Holmes [70], Köthe [86], and Ray [122] as well as the several original articles cited in the text.

Exercises

1. Let $f : I \to \mathbb{R}$ be a continuous function defined on an interval I. If f satisfies $f((x + y)/2) \leq (f(x) + f(y))/2$ for all $x, y \in I$, prove that f is convex.

2. A convex function $f : I \to \mathbb{R}$ is said to be *strictly convex* if $f(\lambda x + (1 - \lambda)y) < \lambda f(x) + (1 - \lambda)f(y)$ whenever $0 < \lambda < 1$ and $x \neq y$. Show that $f(x) = |x|^p$ is strictly convex for $1 < p < \infty$.

3. Give a direct proof that $L_p(\mu)$ is strictly convex.

4. Prove that a normed space X is strictly convex if and only if $\| \cdot \|_X$ is a strictly convex function.

5. Show that every subspace of a strictly convex space is strictly convex. Is the same true for uniformly convex spaces? For smooth spaces?

6. Prove that $\|\|x\|\| = \|x\|_1 + \|x\|_2$ defines an equivalent strictly convex norm on ℓ_1.

7. Suppose that $T : X \to Y$ is continuous and one-to-one and that Y is strictly convex. Show that $\|\|x\|\| = \|x\| + \|Tx\|$ defines an equivalent strictly convex norm on X.

8. Give a direct proof that ℓ_p, $1 < p < \infty$, is strictly convex. Generalize your proof to conclude that if X_n is strictly convex for each n, then $(X_1 \oplus X_2 \oplus \cdots)_p$ is strictly convex for any $1 < p < \infty$.

9. Let $2 \leq p < \infty$ and define $g \in L_p$ by $g(t) = 1$ for $0 \leq t \leq 1 - (\varepsilon/2)^p$, and $g(t) = -1$ for $1 - (\varepsilon/2)^p < t \leq 1$. If we set $f \equiv 1$, show that $\|f\|_p = \|g\|_p = 1$, $\|f - g\|_p = \varepsilon$, and $\|(f + g)/2\|_p = (1 - (\varepsilon/2)^p)^{1/p}$. Conclude that $\delta(\varepsilon) = 1 - (1 - (\varepsilon/2)^p)^{1/p}$ is best possible for L_p.

10. Choose a sequence (p_n) with $1 < p_n < \infty$ such that $p_n \to 1$ or $p_n \to \infty$ and define $X_n = \ell_{p_n}^n$. Show that $(X_1 \oplus X_2 \oplus \cdots)_2$ is strictly convex (and even reflexive) but not uniformly convex.

11. If $x_n \xrightarrow{w} x$ in X, show that $\|x\| \leq \liminf_{n \to \infty} \|x_n\|$. Give an example where this inequality is strict. If $\|x_n\| \leq \|x\|$ for all n, conclude that we actually have $\|x\| = \lim_{n \to \infty} \|x_n\|$. Show that the same holds for a sequence $x_n^* \xrightarrow{w^*} x^*$ in X^*.

12. Apply McShane's lemma to show that each linear map $T : \mathbb{R}^n \to \mathbb{R}$ is given by inner product against a vector $x \in \mathbb{R}^n$ with $\|x\|_2 = \|T\|$.

13. Let $x \in \ell_1$ with $\|x\|_1 \leq 1$. Show that x is an exposed point of the closed unit ball of ℓ_1 if and only if $x = e_k$ for some k.

14. Let $f, (f_n)$ be in L_p, $0 < p < \infty$. If $f_n \to f$ a.e. and $\|f_n\|_p \to \|f\|_p$, show that $\|f_n - f\|_p \to 0$. [Hint: Use the fact that $2^p(|f_n|^p + |f|^p) - |f_n - f|^p$ is nonnegative and tends pointwise a.e. to $2^{p+1}|f|^p$.] Show that the result also holds if we assume instead that $f_n \to f$ in measure.

15. (a) Show that a point x in the closed unit ball of ℓ_∞^n is an extreme point if and only if $x = (\varepsilon_1, \ldots, \varepsilon_n)$, where $\varepsilon_i = \pm 1$ for $i = 1, \ldots, n$. Prove that each point in the unit ball of ℓ_∞^n can be written as a convex combination of extreme points.

    (b) Show that the set of extreme points of the unit ball of ℓ_∞ consists of all points of the form (ε_n), where $\varepsilon_i = \pm 1$ for all $i = 1, 2, \ldots$. Show that the set of extreme points of the unit ball of ℓ_1 consists of the points $\pm e_k$, $k = 1, 2, \ldots$. For $1 < p < \infty$, every norm one vector in ℓ_p is an extreme point of the unit ball.

    (c) In sharp contrast to the previous cases, show that the closed unit ball in c_0 has *no* extreme points.

Chapter 12
$C(K)$ Spaces

The Cantor Set

In this chapter we will catalog several important properties of the Cantor set Δ. Our goal in this endeavor is to uncover the "universal" nature of $C(\Delta)$. For starters, we'll prove that Δ is the "biggest" compact metric space by showing that every compact metric space K is the continuous image of Δ. Now, the presence of a continuous onto map $\varphi : \Delta \to K$ tells us something about $C(K)$. Composition with φ, that is, the map $f \mapsto f \circ \varphi$, defines a linear isometry (and an algebra isomorphism) from $C(K)$ into $C(\Delta)$. Since Δ is itself a compact metric space, this means that $C(\Delta)$ is "biggest" among the spaces $C(K)$ for K compact metric.

To begin, recall that the each element of the Cantor set has a ternary (base 3) decimal expansion of the form $\sum_{n=1}^{\infty} a_n/3^n$, where $a_n = 0$ or 2, and that the Cantor function $\varphi(\sum_{n=1}^{\infty} a_n/3^n) = \sum_{n=1}^{\infty} a_n/2^{n+1}$ defines a continuous map from Δ onto $[0, 1]$. This proves

Lemma 12.1. *The interval* $[0, 1]$ *is the continuous image of* Δ.

As an immediate corollary, we have that $C[0, 1]$ is isometric to a closed subspace (and subalgebra) of $C(\Delta)$. Does this sound backwards? If so, have patience!

For our purposes, a convenient representation of the Cantor set is as the countable product of two-point discrete spaces. This representation is easy to derive from the ternary decimal representation of Δ. That is, since the elements of Δ are sequences of 0s and 2s, we have $\Delta = \{0, 2\}^{\mathbb{N}}$. This representation also yields a natural metric on Δ: Setting $d(x, y) = \sum_{n=1}^{\infty} |a_n - b_n|/3^n$, where a_n, $b_n = 0, 2$ are the ternary decimal digits for $x, y \in \Delta$, respectively, defines a metric equivalent to the usual metric on Δ. (That is, Δ is homeomorphic to the product space $\{0, 2\}^{\mathbb{N}}$ supplied with the metric d.) Moreover, this particular metric has the additional property that $d(x, y) = d(x, z)$ now implies that

$y = z$. Whenever we speak of "the" metric on Δ, this will be the one we have in mind.

Many authors take $\Delta = \{-1, 1\}^{\mathbb{N}}$ since this choice makes Δ a *group* under (coordinatewise) multiplication.

Now since $\mathbb{N}$ can be partitioned into countably many, countably infinite subsets, the product space representation of Δ yields an immediate improvement on our first lemma.

Corollary 12.2. *The cube* $[0, 1]^{\mathbb{N}}$ *is the continuous image of* Δ.

Again, this means that $C([0, 1]^{\mathbb{N}})$ is isometric to a closed subspace of $C(\Delta)$.

Completely Regular Spaces

Recall that a topological space X is said to be *completely regular* if X is Hausdorff and if, given a point $x \in X$ outside a closed set $F \subset X$, there is a continuous function $f : X \to [0, 1]$ such that $f(x) = 1$ while $f = 0$ on F. In other words, X is completely regular if $C(X; [0, 1])$ *separates points from closed sets* in X. Since singletons are closed in X, it's immediate that $C(X; [0, 1])$ also *separates points* in X.

From Urysohn's lemma, every normal space is completely regular. Thus, metric spaces and compact Hausdorff spaces are completely regular. Locally compact Hausdorff spaces are completely regular, too. In fact, nearly every topological space encountered in analysis is completely regular – even topological groups and topological vector spaces. From our point of view, there's no harm in simply assuming that *every* topological space is completely regular.

As it happens, the class of completely regular spaces is precisely the class of spaces that embed in some cube $[0, 1]^A$. To see this, we start with

Lemma 12.3 (Embedding Lemma). *Let X be completely regular. Then X is homeomorphic to a subset of the cube* $[0, 1]^{C(X; [0,1])}$.

Proof. For simplicity, let's write $C = C(X; [0, 1])$. Let $e : X \to [0, 1]^C$ be the evaluation map defined by $e(x)(f) = f(x)$ for $x \in X$ and $f \in C$. That is, x is made to correspond to the "tuple" $e(x) = (f(x))_{f \in C}$. That e is one-to-one is obvious because C separates points in X. That e is continuous is easy, too, for each of its "coordinates" $\pi_f \circ e = f$, $f \in C$, is continuous.

Now let U be open in X. We want to show that $e(U)$ is open in $e(X)$. Here's where we need complete regularity: Given $x \in U$, choose an $f \in C$ such that

$f(x) = 1$ while $f = 0$ on U^c and let $V = \pi_f^{-1}(\{0\}^c)$. Clearly, V is open in the product $[0, 1]^C$ and $e(x) \in V$ because $e(x)(f) = 1 \neq 0$. Finally, $e(y) \in V \implies f(y) \neq 0 \implies y \in U$ and hence $e(x) \in V \cap e(X) \subset e(U)$. $\square$

Our proof of the Embedding lemma shows that if $\mathcal{F} \subset C(X; [0, 1])$ and if $e : X \to [0, 1]^{\mathcal{F}}$ is defined by $e(x)(f) = f(x)$, then e is continuous. If, in addition, $\mathcal{F}$ separates points, then e is one-to-one. Finally, if $\mathcal{F}$ separates points from closed sets, then e is a homeomorphism (into).

The Embedding lemma also tells us that X carries the weak topology induced by $C(X; [0, 1])$. In terms of nets,

$\quad x_\alpha \to x$ in X

$\qquad \Longleftrightarrow e(x_\alpha) \to e(x)$ in $[0, 1]^C$

$\qquad \Longleftrightarrow \pi_f(e(x_\alpha)) \to \pi_f(e(x))$ for every $f \in C(X; [0, 1])$

$\qquad \Longleftrightarrow f(x_\alpha) \to f(x)$ for every $f \in C(X; [0, 1])$

$\qquad \Longleftrightarrow x_\alpha \to x$ in the weak topology induced by $C(X; [0, 1])$.

Let's consolidate several of these observations.

Theorem 12.4. *For a Hausdorff topological space X, the following are equivalent:*

(a) *X is completely regular;*
(b) *X embeds in a cube;*
(c) *X embeds in a compact Hausdorff space;*
(d) *X has the weak topology induced by $C(X; [0, 1])$;*
(e) *X has the weak topology induced by $C_b(X)$.*

Proof. The Embedding lemma shows that (a) implies (b) implies (d). Of course, (b) and Tychonoff's theorem imply (c). Since compact Hausdorff spaces are completely regular, and because every subspace of a completely regular space is again completely regular, (c) implies (a).

Next, let's check that (d) implies (a). Suppose that X has the weak topology induced by $C(X; [0, 1])$ and let $x \in X$ be a point outside the closed set $F \subset X$. Then, for some $f_1, \ldots, f_n \in C(X; [0, 1])$ and some $\varepsilon > 0$, the basic open set $U = N(x; f_1, \ldots, f_n; \varepsilon)$ is contained in F^c. That is,

$\quad U = N(x; f_1, \ldots, f_n; \varepsilon)$

$\qquad = \{y \in X : |f_i(y) - f_i(x)| < \varepsilon, \ i = 1, \ldots, n\} \subset F^c$.

Now each of the functions $g_i = |f_i - f_i(x)|$ is in $C(X; [0, 1])$, as is the function $g = \max\{g_1, \ldots, g_n\}$. Moreover, $g(x) = 0$ while $g(y) \geq \varepsilon$ for all

$y \in F \subset U^c$. Hence, $h = \varepsilon^{-1} \min\{\varepsilon, g\} \in C(X; [0, 1])$ satisfies $h(x) = 0$ and $h(y) = 1$ for all $y \in F \subset U^c$. Thus, X is completely regular.

Finally, the fact that (d) is equivalent to (e) is easy. Since $C(X; [0, 1]) \subset C_b(X)$, it's clear that each "$C(X; [0, 1])$ open" set is also "$C_b(X)$ open." On the other hand, given $x \in X$, $f \in C_b(X)$, and $\varepsilon > 0$, put $M = \|f\|_\infty$ and notice that $g = (f + M)/2M \in C(X; [0, 1])$ satisfies

$$\{y \in X : |f(x) - f(y)| < \varepsilon\} = \{y \in X : |g(x) - g(y)| < \varepsilon/2M\}.$$

Thus, each "$C_b(X)$ open" set is also "$C(X; [0, 1])$ open." □

If X is completely regular and if some *countable* family $\mathcal{F} \subset C(X; [0, 1])$ separates points from closed sets in X, then X embeds in the cube $[0, 1]^{\mathbb{N}}$ and hence is metrizable. (Compare this with Urysohn's metrization theorem: A normal, second countable space is metrizable.) This result has a converse of sorts, too. If, for example, X is a *separable* metric space, then some countable family $\mathcal{F} \subset C(X; [0, 1])$ will separate points from closed sets in X. In particular, X will embed in $[0, 1]^{\mathbb{N}}$. More to the point for us is

Lemma 12.5. *Every compact metric space is homeomorphic to a closed subset of* $[0, 1]^{\mathbb{N}}$.

Proof. Let K be a compact metric space and let (x_n) be dense in K. We may assume that the metric d on K satisfies $d(x, y) \leq 1$. Given this, define $\psi :$ $K \to [0, 1]^{\mathbb{N}}$ by $\psi(x)(n) = d(x, x_n)$. Clearly, ψ is one-to-one and continuous (each coordinate $\psi(\cdot)(n) = d(\cdot, x_n)$ is continuous). Since K is compact and $[0, 1]^{\mathbb{N}}$ is Hausdorff, ψ is a homeomorphism (into) and the result follows. □

Lest you be fooled into thinking that only countable products of intervals are separable, here is an easy counterexample:

Example. $[0, 1]^{[0,1]}$ is separable but not metrizable.

Proof. Consider the collection $\mathcal{D}$ of all functions of the form $\sum_{i=1}^n q_i \chi_{J_i}$, where $q_1, \ldots, q_n$ are rationals in $[0, 1]$ and where $J_1, \ldots, J_n$ are disjoint closed intervals in $[0, 1]$ with rational endpoints. Then $\mathcal{D}$ is a countable subset of $[0, 1]^{[0,1]}$.

Now the typical basic open set in $[0, 1]^{[0,1]}$ is given by

$$N(f; x_1, \ldots, x_n; \varepsilon) = \{g : |g(x_i) - f(x_i)| < \varepsilon, \ i = 1, \ldots, n\},$$

where $x_1, \ldots, x_n \in [0, 1]$, $\varepsilon > 0$, and $f : [0, 1] \to [0, 1]$. Given a basic open set $N(f; x_1, \ldots, x_n; \varepsilon)$, choose rationals $q_1 \ldots, q_n$ in $[0, 1]$ with $|q_i - f(x_i)| < \varepsilon$ for each i, and choose disjoint rational intervals J_i with $x_i \in J_i$. Then $g = \sum_{i=1}^{n} q_i \chi_{J_i} \in \mathcal{D} \cap N(f; x_1, \ldots, x_n; \varepsilon)$. Thus, $\mathcal{D}$ is dense, and so $[0, 1]^{[0,1]}$ is separable.

To see that $[0, 1]^{[0,1]}$ is not metrizable, note that it's not sequentially compact. Your favorite sequence of functions $f_n : [0, 1] \to [0, 1]$ with no pointwise convergent subsequence will do the trick. The sequence $(1 + r_n)/2$, where r_n is the nth Rademacher function, comes to mind. $\quad\square$

We're now ready to establish our claim that the Cantor set is the "biggest" compact metric space.

Theorem 12.6. *Every compact metric space K is the continuous image of Δ.*

Proof. We know that K is homeomorphic to a closed subset of $[0, 1]^{\mathbb{N}}$ and, hence, that K is the continuous image of some closed subset of Δ. To finish the proof, then, it suffices to show that each closed subset of Δ is the continuous image of Δ.

So, let F be a closed subset of Δ and let $d(x, y) = \sum_{n=1}^{\infty} |a_n - b_n|/3^n$ be "the" metric on Δ. Given $x \in \Delta$, notice that the distance $d(x, F) = \inf_{y \in F} d(x, y)$ is attained at a unique point $y \in F$ (and, in particular, $y = x$ whenever $x \in F$). Define $f(x) = y$ for this unique y. To check that f is continuous, let $x_n \to x$ in Δ. Now, since F is compact, we may assume that $y_n = f(x_n)$ converges to some point $z \in \Delta$. But then $d(x_n, y_n) \to d(x, z)$ and $d(x_n, y_n) = d(x_n, F) \to d(x, F) = d(x, y)$, and so we must have $y = z$; that is, $y_n \to y = f(x)$. $\quad\square$

Corollary 12.7. *If K is a compact metric space, then $C(K)$ is isometric to a closed subspace (even a subalgebra) of $C(\Delta)$.*

This completes the first circle of ideas in this chapter: $C(\Delta)$ is "universal" for the class of spaces $C(K)$, K compact metric. This may seem odd in view of the small role that Δ typically plays in a first course in analysis. Indeed, $C[0, 1]$ is usually given much more emphasis. The reader who is uncomfortable with this turn of fate can take heart from the fact that $C(\Delta)$ embeds isometrically into $C[0, 1]$. That is, $C[0, 1]$ is universal, too. We'll need a bit more machinery before we can give a proof – note that we can't hope to prove this claim by finding a continuous map from $[0, 1]$ onto Δ!

Now each $f \in C(\Delta)$ extends to a continuous function on $[0, 1]$. This would follow from Tietze's extension theorem, for example, but there may be many such extensions. We want to choose a method for extending f that will lead to a *linear* map from $C(\Delta)$ into $C[0, 1]$. And the most natural extension does the job.

The complement of Δ in $[0, 1]$ is the countable union of disjoint open intervals. The endpoints of these open intervals are, of course, elements of Δ. Simply "connect the dots" in the graph of f across the endpoints of each of these open intervals to define an extension $\tilde{f}$ for f. (This is quite like the procedure used to extend the Cantor function to all of $[0, 1]$.) Note that if $x \notin \Delta$, then $\tilde{f}(x)$ is the "average" of two values of f on Δ. Clearly, then, $\sup_{0 \leq x \leq 1} |\tilde{f}(x)| = \sup_{x \in \Delta} |f(x)|$. What's more, it's not hard to see that if we're given $f, g \in C(\Delta)$, then $\widetilde{f + g} = \tilde{f} + \tilde{g}$. We'll take this as proof of

Lemma 12.8. *The extension map* $E(f) = \tilde{f}$ *from* $C(\Delta)$ *into* $C[0, 1]$ *is a linear isometry.*

Theorem 12.9. $C(\Delta)$ *is isometric to a complemented subspace of* $C[0, 1]$.

Proof. To complete the proof, we need only note that restriction to Δ defines a linear map $R : C[0, 1] \to C(\Delta)$ with the property that $P = ER$ is the identity on $E(C(\Delta))$. That is, $E(C(\Delta))$ is an isometric copy of $C(\Delta)$ and is the range of a bounded projection $P : C[0, 1] \to C[0, 1]$. $\square$

Corollary 12.10. *If* K *is a compact metric space, then* $C(K)$ *is isometric to a closed subspace of* $C[0, 1]$. *In particular,* $C(K)$ *is separable.*

Our last corollary should be viewed as a rough analogue of the Weierstrass theorem valid in $C(K)$ for a compact metric space K. What's more, the converse is also true: If $C(K)$ is separable for a given compact Hausdorff space K, then K is metrizable. We'll forego the details just now, but this issue will come up again.

The universality of $C(\Delta)$ (or $C[0, 1]$) reaches beyond the class of $C(K)$ spaces; in fact, *every* separable normed space is isometric to a subspace of $C(\Delta)$. (Even more is true: Every separable *metric* space is isometric to a subset of $C(\Delta)$.) By our last corollary, we only need to check that each separable normed linear space embeds in some $C(K)$, K compact metric. At least part of this claim is easy to check and is valid for any normed linear space. As a consequence of the Hahn–Banach theorem, a normed linear space X is isometric to a subspace of $C_b(B_{X^*})$, under the sup norm, via point evaluation: $x \mapsto \hat{x}|_{B_{X^*}}$, where $\hat{x}(x^*) = x^*(x)$. That is, the embedding of X into $C_b(B_{X^*})$ is nothing other than the canonical embedding of X into X^{**} with the additional restriction that each element of X^{**} is to be considered as a function just on B_{X^*}.

What remains to be seen is whether B_{X^*} can be given a suitable topology that will turn it into a compact metric space (thus making $C_b(B_{X^*}) = C(B_{X^*})$). For an infinite-dimensional space X, the norm topology on B_{X^*} won't do (it's never compact). We'll need something weaker. The topology that fits the bill is the weak* topology on X^* restricted to B_{X^*}.

The weak* topology is the topology that X^* inherits as a subset of the product space $\mathbb{R}^X$. That is, each $x^* \in X^*$ is identified with its range $(x^*(x))_{x \in X} = (\hat{x}(x^*))_{x \in X}$ considered as an element of $\mathbb{R}^X$. Under this identification, $\hat{x}$ is the projection onto the "xth coordinate" of X^*. Thus, the weak* topology on X^* is the smallest topology on X^* making every $\hat{x}$ continuous (or, in still other words, the weak* topology is the weak topology on X^* induced by $\widehat{X} = X$). A neighborhood base for the weak* topology is generated by the sets

$$N(x^*; x_1, \ldots, x_k; \varepsilon) = \{y^* \in X^* : |(y^* - x^*)(x_i)| < \varepsilon \text{ for } i = 1, \ldots, k\},$$

where $x^* \in X^*$, $x_1, \ldots, x_k \in X$, and $\varepsilon > 0$. In terms of nets,

$$x_\alpha^* \xrightarrow{w^*} x^* \text{ in } X^* \iff \hat{x}(x_\alpha^*) \to \hat{x}(x^*) \text{ for each } \hat{x} \in \hat{X}$$
$$\iff x_\alpha^*(x) \to x^*(x) \text{ for each } x \in X$$
$$\iff x_\alpha^* \to x^* \text{ in } \mathbb{R}^X.$$

Of particular merit here is that X is still isometric to a subspace of $C_b(B_{X^*})$ even when B_{X^*} is supplied with the weak* topology (since each $\hat{x}$ is weak*-continuous). That this observation has brought us one step closer to our goal is given as

Theorem 12.11 (Banach–Alaoglu). *If X is a normed linear space, then B_{X^*} is compact in the weak* topology on X^*.*

Proof. As we've already seen, X^*, under the weak* topology, is homeomorphic to a subset of $\mathbb{R}^X$. If we cut down to B_{X^*}, then we can do a bit better. Again we identify each $x^* \in B_{X^*}$ with its range $(x^*(x))_{x \in X}$ but now considered as an element of the product $\prod_{x \in X}[-\|x\|, \|x\|]$. This identification is still a homeomorphism (into) by virtue of the definition of the weak* topology. Since $\prod_{x \in X}[-\|x\|, \|x\|]$ is compact, we will be done once we show that the image of B_{X^*} is closed in the product topology. But what does this really mean? We need to check that the pointwise limit of a net $x_\alpha^* \in B_{X^*}$ of linear functions on X is again linear and has norm at most one – which is clear. $\square$

Corollary 12.12. *Every normed linear space is isometric to a subspace of* $C(K)$, *for some compact Hausdorff space* K.

We want to embed a separable normed space X into $C(K)$, where K is a compact *metric* space. Our next result shows the way.

Theorem 12.13. *If* X *is a separable normed linear space, then the weak* topology on* B_{X^*} *is both compact and metrizable.*

Proof. That B_{X^*} is compact in the weak* topology is immediate. Now let (x_n) be dense in S_X and define $d(x^*, y^*) = \sum_{n=1}^{\infty} |(x^* - y^*)(x_n)|/2^n$ for x^*, $y^* \in B_{X^*}$. It's easy to see that d is a metric on B_{X^*}. Next, notice that

$$d(x^*, y^*) \leq \max_{1 \leq n \leq M} |(x^* - y^*)(x_n)| \sum_{n=1}^{M} 2^{-n} + 2 \sum_{n=M+1}^{\infty} 2^{-n}$$

$$< \max_{1 \leq n \leq M} |(x^* - y^*)(x_n)| + 2^{-M+1}$$

because $\|x^* - y^*\| \leq 2$ for x^*, $y^* \in B_{X^*}$. Here's what this does: Given $x^* \in B_{X^*}$ and $\varepsilon > 0$, if M is chosen so that $2^{-M+1} < \varepsilon/2$, then

$$N(x^*; x_1, \ldots, x_N; \varepsilon/2) \cap B_{X^*} \subset \{y^* \in B_{X^*} : d(x^*, y^*) < \varepsilon\}.$$

That is, the formal identity from (B_{X^*}, weak^*) to (B_{X^*}, d) is continuous. But since (B_{X^*}, weak^*) is compact and (B_{X^*}, d) is Hausdorff, the two spaces must actually be homeomorphic. $\square$

Corollary 12.14 (Banach–Mazur). *Every separable normed linear space is isometric to a subspace of* $C(K)$ *for some compact metric space* K. *Thus, every separable normed linear space is isometric to a subspace of* $C[0, 1]$.

If follows from the Banach–Mazur theorem that $C[0, 1]$ isn't reflexive since it contains an isometric copy of the nonreflexive space ℓ_1. In fact, this same observation tells us that $C[0, 1]^*$ must be nonseparable.

Corollary 12.15. *Every separable metric space is isometric to a subset of* $C(\Delta)$.

Proof. Suppose that (M, d) is a separable metric space and, let (x_n) be dense in (M, d). Fix a point $x_0 \in M$ and define a map from M into ℓ_∞ by

$$x \longmapsto \tilde{x} = (d(x, x_n) - d(x_0, x_n))_{n=1}^{\infty} .$$

By the triangle inequality in (M, d), we have $\tilde{x} \in \ell_\infty$ and $\|\tilde{x}\|_\infty \le d(x, x_0)$. Essentially the same observation shows that the map is an isometry:

$$\|\tilde{x} - \tilde{y}\|_\infty = \sup_n |d(x, x_n) - d(y, x_n)| = d(x, y).$$

Indeed, it's clear from the triangle inequality in (M, d) that $\|\tilde{x} - \tilde{y}\|_\infty \le d(x, y)$. On the other hand, taking x_n with $d(y, x_n) < \varepsilon$ gives $\|\tilde{x} - \tilde{y}\|_\infty \ge |d(x, x_n) - d(y, x_n)| \ge d(x, y) - 2\varepsilon$. Thus, M is isometric to a subset $\tilde{M}$ of ℓ_∞. In particular, the sequence $(\tilde{x}_n)$ in ℓ_∞ is dense in $\tilde{M}$. It follows that the closed linear span of $\tilde{M}$ is a *separable* subspace of ℓ_∞. An appeal to the Banach–Mazur theorem finishes the proof. $\square$

Here is our long-awaited application of the Banach–Mazur theorem.

Corollary 12.16. *Every normed linear space contains an infinite-dimensional closed subspace with a basis.*

Finally we're ready to complete a second circle of ideas from this chapter: The separability of $C(K)$ is equivalent to the metrizability of K. We consider the (apparently) more general case of a completely regular space T (to save wear and tear on the letter X).

Theorem 12.17. *If T is a completely regular topological space, then T is homeomorphic to a subset of $(B_{C_b(T)^*}, \text{weak}^*)$.*

Proof. Each point $t \in T$ induces an element $\delta_t \in C_b(T)^*$ by way of (what else!) point evaluation: $\delta_t(f) = f(t)$. [The functional δ_t is the *point mass* or *Dirac measure* at t.] Note that δ_t is norm one; hence, $\delta_t \in B_{C_b(T)^*}$. Moreover, it follows easily that the map $t \mapsto \delta_t$ is actually a homeomorphism from T into $(B_{C_b(T)^*}, \text{weak}^*)$. The key is that the completely regular space T carries the weak topology induced by $C_b(T)$. In terms of nets,

$$\begin{aligned}
t_\alpha \to t \text{ in } T \iff & f(t_\alpha) \to f(t) \text{ for all } f \in C_b(T) \\
\iff & \delta_{t_\alpha}(f) \to \delta_t(f) \text{ for all } f \in C_b(T) \\
\iff & \delta_{t_\alpha} \xrightarrow{w^*} \delta_t. \quad \square
\end{aligned}$$

Corollary 12.18. *Let T be completely regular. If $C_b(T)$ is separable, then T is metrizable.*

Proof. Since $C_b(T)$ is separable, we get that $(B_{C_b(T)^*}, \text{weak}^*)$ is metrizable. By our previous result, T is then metrizable too. $\square$

Corollary 12.19. *Let K be a compact Hausdorff space. Then, $C(K)$ is separable if and only if K is metrizable.*

Before we leave these topics, we would be wise to say a few words about the weak* topology on X^{**}. The weak* topology on X^{**} is the topology X^{**} inherits as a subspace of $\mathbb{R}^{X^*}$; that is, the smallest topology on X^{**} making each element of X^* continuous. In terms of nets,

$$x_\alpha^{**} \xrightarrow{w^*} x^{**} \iff x_\alpha^{**}(x^*) \to x^{**}(x^*) \text{ for all } x^* \in X^*.$$

A neighborhood base for the weak* topology is given by the sets

$$N(x^{**}; x_1^*, \ldots, x_n^*; \varepsilon)$$
$$= \{y^{**} \in X^{**} : |y^{**}(x_i^*) - x^{**}(x_i^*)| < \varepsilon, \ i = 1, \ldots, n\},$$

where $x^{**} \in X^{**}$, $x_1^*, \ldots, x_n^* \in X^*$, and $\varepsilon > 0$.

As a first observation, notice that the weak* topology on X^{**}, when restricted to $\widehat{X}$, is nothing but the weak topology on X. In terms of nets,

$$\hat{x}_\alpha \xrightarrow{w^*} \hat{x} \text{ in } \widehat{X} \iff \hat{x}_\alpha(x^*) \to \hat{x}(x^*) \text{ for all } x^* \in X^*$$
$$\iff x^*(x_\alpha) \to x^*(x) \text{ for all } x^* \in X^*$$
$$\iff x_\alpha \xrightarrow{w} x \text{ in } X.$$

Thus, $(\widehat{X}, \text{weak}^*)$ is homeomorphic to (X, weak) or, in other words, the map $x \mapsto \hat{x}$ is a weak-to-weak* homeomorphism from X into X^{**}.

Next, from the Banach–Alaoglu theorem, we know that $B_{X^{**}}$ is compact in the weak* topology on X^{**}. Hence, if X is reflexive, then B_X is weakly compact in X. The converse is also true: If B_X is weakly compact in X, then X is reflexive. Indeed, in this case, we would have that $B_{\widehat{X}}$ is weak* compact and weak* dense in $B_{X^{**}}$ (from Goldstine's theorem). Thus, $B_{\widehat{X}} = B_{X^{**}}$, and it follows that $\widehat{X} = X^{**}$. We take this as proof of

Proposition 12.20. *X is reflexive if and only if B_X is weakly compact.*

(Compare this result to the fact that X is finite dimensional if and only if B_X is norm compact.)

Notes and Remarks

Our presentation in this chapter borrows heavily from Lacey [88], but see also Folland [48, Chapter 4], Kelley [84], and Willard [146]. Theorem 12.6 is semi-classical; Kelley attributes the result to Alexandroff and Urysohn (from 1923, I believe) but gives an ambiguous reference; Banach refers to the 1927 edition of Hausdorff's book [see 66]. The Banach–Alaoglu theorem (Theorem 12.11) was first proved by Banach [6] for separable spaces and later improved to its present state by Alaoglu [2]. Alaoglu's paper also contains some interesting applications, including some contributions to the problem of embedding normed spaces into $C(K)$ spaces (along the lines of Corollary 12.15). The Banach–Mazur theorem (Theorem 12.14) appears in Banach's book [6].

Exercises

1. Let X and Y be Hausdorff topological spaces with X compact. If $f : X \to Y$ is one-to-one and continuous, prove that f is a homeomorphism (into).

2. Let d be the metric on Δ defined by $d(x, y) = \sum_{n=1}^{\infty} |a_n - b_n|/3^n$, where $a_n, b_n = 0, 2$ are the ternary decimal digits for $x, y \in \Delta$.
    (a) If $d(x, y) < 3^{-n}$, show that the first n ternary digits of x and y agree.
    (b) Show that d is equivalent to the usual metric on Δ. [Hint: The identity map from (Δ, d) to $(\Delta,$ usual$)$ is continuous.]
    (c) Show that $d(x, y) = d(x, z)$ implies $y = z$.

3. Prove Corollary 12.2.

4. If X is a separable metric space, prove that some countable family $\mathcal{F} \subset C(X; [0, 1])$ will separate points from closed sets in X.

5. Prove Lemma 12.8, filling in any missing details about the extension operator E.

6. Show that every subspace of a completely regular space is completely regular.

7. (a) If A is a subset of X^*, show that $^{\perp}A$ is a closed subspace of X.
    (b) If A is a subset of X, show $A^{\perp}$ is a weak* closed subspace of X^*.
    (c) If A is a subset of X, show that $A^{\perp} = {}^{\perp}(\widehat{A})$, where $\widehat{A}$ is the canonical image of A in X^{**}.

(d) If Y is a subspace of X^{**}, show that $(^{\perp}Y)^{\perp}$ is the weak* closure of Y in X^{**}. If Y is a subspace of X, conclude that $Y^{\perp\perp}$ is the weak* closure of $\widehat{Y}$ in X^{**}.

(e) (Goldstine's theorem, junior grade): Prove that $\widehat{X}$ is weak* dense in X^{**}.

8. (Goldstine's theorem, utility grade): Prove that $B_{\widehat{X}}$ is weak* dense in $B_{X^{**}}$. [Hint: If $F \in B_{X^{**}}$ is outside the weak* closure of $B_{\widehat{X}}$, then there is a weak* continuous linear functional separating F from $B_{\widehat{X}}$.]

Chapter 13

Weak Compactness in L_1

In this chapter we examine the structure of weakly compact subsets of $L_1 = L_1[0, 1]$. Since the closed unit ball of any reflexive space is weakly compact, this undertaking will lead to a better understanding of the reflexive subspaces of L_1. It should be pointed out that many of our results will hold equally well in $L_1(\mu)$, where μ is a finite measure.

We begin with a definition: We say that a subset F of L_1 is *uniformly integrable* (or, as some authors say, *equi-integrable*) if

$$\sup_{f \in F} \int_{\{|f| > a\}} |f(t)| dt \to 0 \quad \text{as} \quad a \to \infty.$$

What this means is that all of the elements of F can be truncated at height a with uniform error (in the L_1 norm). A few examples might help.

Examples

1. Given $f \in L_1$, it's an easy consequence of Chebyshev's inequality that $m\{|f| > a\} \searrow 0$ as $a \nearrow \infty$. Thus, since $A \mapsto \int_A |f|$ is absolutely continuous with respect to m, we get

   $$\int_{\{|f| > a\}} |f| \to 0 \quad \text{as} \quad a \to \infty.$$

   In other words, any singleton $\{f\}$ is a uniformly integrable set in L_1. In fact, any finite subset of L_1 is uniformly integrable.

2. If there exists an element $g \in L_1$ such that $|f| \leq |g|$ for all $f \in F$, then F is uniformly integrable. Indeed, given $f \in F$ and $a \in \mathbb{R}$, we would have $\{|f| > a\} \subset \{|g| > a\}$ and, hence,

   $$\sup_{f \in F} \int_{\{|f| > a\}} |f(t)| dt \leq \int_{\{|g| > a\}} |g(t)| dt \to 0.$$

   What this means is that the *order interval*

   $$[-g, g] = \{f : -g \leq f \leq g\}$$

136

is uniformly integrable. It's not hard to see that any order interval $[g, h]$ in L_1 is likewise uniformly integrable.

3. The sequence $f_n = n\chi_{[0,1/n]}$ is *not* uniformly integrable. (Why?)

4. Note that a uniformly integrable subset F of L_1 is necessarily norm bounded. Indeed, if we choose $a \in \mathbb{R}$ such that $\int_{\{|f|>a\}} |f| \leq 1$ for all $f \in F$, then $\|f\|_1 = \int |f| \leq a + 1$ for all $f \in F$.

As our first example might suggest, F is uniformly integrable whenever the family of measures $\{\int_A |f| : f \in F\}$ is *uniformly absolutely continuous* with respect to m (or, as some authors say, *equiabsolutely continuous*, or simply *equicontinuous*). This is the content of our first result.

Proposition 13.1. *A subset F of L_1 is uniformly integrable if and only if the family of measures $\mathcal{F} = \{\int_A |f| : f \in F\}$ is uniformly absolutely continuous with respect to m; that is, if and only if, for every $\varepsilon > 0$, there exists a $\delta > 0$ such that $\int_A |f| < \varepsilon$ for all $f \in F$ and for all Borel sets $A \subset [0, 1]$ with $m(A) < \delta$.*

Proof. First suppose that F is uniformly integrable. Given $\varepsilon > 0$, choose a such that $\int_{\{|f|>a\}} |f| < \varepsilon/2$ for all $f \in F$ and now choose $\delta > 0$ such that $a\delta < \varepsilon/2$. Then, for $f \in F$ and $A \subset [0, 1]$, we have

$$\int_A |f| = \int_{A \cap \{|f|>a\}} |f| + \int_{A \cap \{|f| \leq a\}} |f|$$
$$\leq \varepsilon/2 + a \cdot m(A) < \varepsilon$$

whenever $m(A) < \delta$. Thus, $\mathcal{F}$ is uniformly absolutely continuous.

Now suppose that $\mathcal{F}$ is uniformly absolutely continuous. Then F is norm bounded. Indeed, choose $\delta > 0$ such that $\int_A |f| < 1$ whenever $f \in F$ and $m(A) < \delta$, and partition $[0, 1]$ into $M = 1 + [2/\delta]$ subintervals, each of length at most $\delta/2$. It follows that $\|f\|_1 < M$ for each $f \in F$.

Given $\varepsilon > 0$, choose $\delta > 0$ such that $\int_A |f| < \varepsilon$ whenever $f \in F$ and $m(A) < \delta$. Next, from Chebyshev's inequality, notice that for $f \in F$ we have

$$m\{|f| > a\} \leq \frac{\|f\|_1}{a} < \frac{M}{a}.$$

Thus, if a is chosen so that $M/a < \delta$, then $\int_{\{|f|>a\}} |f| < \varepsilon$. Thus, F is uniformly integrable. $\square$

To place our discussion in the proper context, it will help to recall the *measure algebra* associated with $([0, 1], \mathcal{B}, m)$. To begin, it is an easy exercise to check that $d(A, B) = m(A \triangle B)$ defines a pseudometric on $\mathcal{B}$. Thus, if we define an equivalence relation on $\mathcal{B}$ by declaring $A \sim B$ if and only if $m(A \triangle B) = 0$, and if we write $\widetilde{\mathcal{B}}$ to denote the set of equivalence classes under this relation, then d induces a metric on $\widetilde{\mathcal{B}}$. In fact, d induces a *complete* metric on $\widetilde{\mathcal{B}}$. This is easiest to see if we first make an observation: For A and B in $\mathcal{B}$, we have

$$m(A \triangle B) = d(A, B) = \| \chi_A - \chi_B \|_1.$$

Thus, if we agree to equate sets that are a.e. equal, then $\widetilde{\mathcal{B}}$ inherits a complete metric from L_1. The complete metric space $(\widetilde{\mathcal{B}}, d)$ is called the measure algebra associated with $([0, 1], \mathcal{B}, m)$.

In this setting, a Borel measure μ is absolutely continuous with respect to m if and only if μ is continuous at $\emptyset$ in $(\widetilde{\mathcal{B}}, d)$. (Measures behave rather like linear functions on $\widetilde{\mathcal{B}}$. In particular, we only need to worry about continuity at "zero.") And a subset F of L_1 is uniformly integrable if and only if the corresponding set of measures $\mathcal{F}$ is *equicontinuous at $\emptyset$* when considered as a collection of functions on $(\widetilde{\mathcal{B}}, d)$.

A special case is worth isolating:

Lemma 13.2. *Given $f \in L_1$, define $\Phi : \widetilde{\mathcal{B}} \to \mathbb{R}$ by $\Phi(A) = \int_A f$. Then, Φ is uniformly continuous on $(\widetilde{\mathcal{B}}, d)$.*

Proof. (Note that Φ is welldefined.) This is a simple computation:

$$\left| \int_A f - \int_B f \right| = \left| \int_{A \setminus B} f - \int_{B \setminus A} f \right| \leq \int_{A \triangle B} |f|.$$

Thus, $|\Phi(A) - \Phi(B)|$ can be made small provided that $d(A, B) = m(A \triangle B)$ is sufficiently small (where "small" depends on f but not on A or B). $\square$

Corollary 13.3. *Given $f \in L_1$ and $\varepsilon > 0$, the set $\{A : |\int_A f| \leq \varepsilon\}$ is closed in $(\widetilde{\mathcal{B}}, d)$.*

Finally we're ready to make some connections with weak compactness (or, in this case, weak convergence) in L_1.

Proposition 13.4. *Let (f_n) be a sequence in L_1 such that $\lim_{n\to\infty}\int_A f_n$ exists in $\mathbb{R}$ for every Borel subset A of $[0, 1]$. Then*

(a) *(f_n) is uniformly integrable and*
(b) *(f_n) converges weakly to some f in L_1; in particular, $\int_A f_n \to \int_A f$ for every Borel subset A of $[0, 1]$.*

Proof. For $\varepsilon > 0$ and $N = 1, 2, \ldots$, define

$$F_N = \left\{ A \in \widetilde{\mathcal{B}} : \left| \int_A (f_m - f_n) \right| \le \varepsilon \text{ for all } m, n \ge N \right\}.$$

Since $(\int_A f_n)$ is Cauchy for each $A \in \widetilde{\mathcal{B}}$, we have $\widetilde{\mathcal{B}} = \bigcup_{N=1}^{\infty} F_N$. But each F_N is closed and $\widetilde{\mathcal{B}}$ is complete. By Baire's theorem, then, some F_{N_0} must have nonempty interior. Thus, there exists some $A_0 \in \widetilde{\mathcal{B}}$ and some $r > 0$ such that $m(A \triangle A_0) < r$ implies $A \in F_{N_0}$.

Suppose now that $m(B) < r$. Then, for $m, n \ge N$, we have

$$\int_B (f_m - f_n) = \int_{A_0 \cup B} (f_m - f_n) - \int_{A_0 \setminus B} (f_m - f_n)$$

and also

$$m((A_0 \cup B)\triangle A_0) < r \quad \text{and} \quad m((A_0 \setminus B)\triangle A_0) < r.$$

Hence, $\left|\int_B (f_m - f_n)\right| \le \varepsilon + \varepsilon = 2\varepsilon$. Applying the same argument to the sets

$$B \cap \{f_m - f_n \ge 0\} \quad \text{and} \quad B \cap \{f_m - f_n \le 0\}$$

then yields $\int_B |f_m - f_n| \le 4\varepsilon$ whenever $m(B) < r$ and $m, n \ge N_0$.

Now the set $F = \{f_j : 1 \le j \le N_0\}$ is uniformly integrable, and so there exists an $s < r$ such that $\int_B |f_j| < \varepsilon$ whenever $1 \le j \le N_0$ and $m(B) < s$. Finally, if $m(B) < s$ and $j > N_0$, we have

$$\int_B |f_j| \le \int_B |f_j - f_{N_0}| + \int_B |f_{N_0}|$$
$$\le 4\varepsilon + \varepsilon = 5\varepsilon.$$

Thus (f_n) is uniformly integrable, which proves (a).

Next, for each $A \in \widetilde{\mathcal{B}}$, put $\Phi(A) = \lim_{n\to\infty}\int_A f_n$. Then Φ is countably additive (this follows from the fact that (f_n) is uniformly integrable) and absolutely continuous with respect to m. Thus, by the Radon–Nikodým theorem, there exists an $f \in L_1$ such that $\Phi(A) = \int_A f$. That is, $\int_A f_n \to \int_A f$ for every $A \in \mathcal{B}$. It follows (from linearity of the integral) that $\int f_n g \to \int fg$ for every simple function g. Since the simple functions are dense in L_∞, we get $\int f_n g \to \int fg$ for every $g \in L_\infty$. That is, $f_n \xrightarrow{w} f$ in L_1, which proves (b). $\square$

Corollary 13.5. *L_1 is weakly sequentially complete.*

Proof. Suppose that (f_n) is weakly Cauchy in L_1. Then, in particular, $\int_A f_n$ converges for every $A \in \mathcal{B}$. Hence, (f_n) converges weakly to some $f \in L_1$ by Proposition 13.4. $\square$

Finally we're ready to characterize the weakly compact subsets of L_1.

Theorem 13.6. *A subset F of L_1 is relatively weakly compact if and only if it is uniformly integrable.*

Proof. Suppose that F is not uniformly integrable. Then there exists an $\varepsilon > 0$ such that for each n we can find an $f_n \in F$ with $\int_{\{|f_n|>n\}} |f_n| \geq \varepsilon$. In particular, (f_n) has no uniformly integrable subsequence. It then follows from Proposition 13.4 that (f_n) has no weakly convergent subsequence. Thus, F cannot be relatively weakly compact. (This follows from the Eberlein–Šmulian theorem: A subset of a normed space is weakly compact if and only if it's weakly sequentially compact.)

Now suppose that F is uniformly integrable. We will show that the weak* closure of F is contained in L_1 (considered as a subset of L_1^{**}) and, hence, that F is weakly compact. To this end, suppose that G is in the weak* closure of F. Given $\varepsilon > 0$, choose $\delta > 0$ such that $\int_A |f| < \varepsilon$ whenever $f \in F$ and $m(A) < \delta$. Since G is a weak* cluster point of F, it follows that $|G(\chi_A)| < \varepsilon$ whenever $m(A) < \delta$. Thus, the measure $A \mapsto G(\chi_A)$ is absolutely continuous with respect to m. By the Radon–Nikodým theorem, there is a $g \in L_1$ such that $G(\chi_A) = \int_A g$ and it now follows (from the linearity and continuity of G) that $G(h) = \int hg$ for every $h \in L_\infty$. That is, $G = g \in L_1$. $\square$

The same proof applies to any $L_1(\mu)$, where μ is a positive finite measure. Thus, if μ is positive and finite, the weakly compact subsets of $L_1(\mu)$ are precisely the uniformly integrable subsets. This leads us to our final (but most useful) corollary.

Corollary 13.7 (Vitali–Hahn–Saks Theorem). *Let (μ_n) be a sequence of signed measures on a σ-algebra Σ such that $\mu(A) = \lim_{n \to \infty} \mu_n(A)$ exists in $\mathbb{R}$ for each $A \in \Sigma$. Then μ is a signed measure on Σ.*

Proof. Let $\lambda = \sum_{n=1}^{\infty} |\mu_n|/2^n \|\mu_n\|$ (where $\|\mu\| = |\mu|(X)$, the total variation of μ applied to the underlying measure space X). Then each μ_n is absolutely continuous with respect to λ; that is, $d\mu_n = f_n \, d\lambda$ for some $f_n \in L_1(\lambda)$. Thus, $\int_A f_n \, d\lambda = \mu_n(A) \to \mu(A)$ for each $A \in \Sigma$, and so, by Proposition 13.4, there exists an $f \in L_1(\lambda)$ such that

$$\int_A f \, d\lambda = \lim_{n \to \infty} \int_A f_n \, d\lambda = \mu(A)$$

for every $A \in \Sigma$. That is, $d\mu = f \, d\lambda$ and it follows that μ is a signed measure (i.e., μ is countably additive) on Σ. $\square$

Notes and Remarks

Our presentation in this chapter borrows heavily from some unpublished notes for a course on Banach space theory given by Stephen Dilworth at the University of South Carolina [35]. The Eberlein–Šmulian theorem can be found in any number of books, but see also [115] and [144].

Exercises

1. If (f_n) converges to f in L_1, give a direct proof that (f_n) is uniformly integrable.

2. Let (f_n) be a sequence in L_1. If $f_n \to f$ a.e., show that the following are equivalent:
   (a) (f_n) is uniformly integrable;
   (b) $\|f_n - f\|_1 \to 0$ as $n \to \infty$;
   (c) $\|f_n\|_1 \to \|f\|_1$ as $n \to \infty$.

3. If (f_n) converges weakly to f in L_1, show that (f_n) is uniformly integrable.

4. Prove that $d(A, B) = m(A \triangle B)$ defines a complete pseudometric on $\mathcal{B}$.

5. Given a measure $\mu : \mathcal{B} \to \mathbb{R}$, prove that the following are equivalent:
   (a) μ is absolutely continuous with respect to m;
   (b) μ is continuous at $\emptyset$ in $(\widetilde{\mathcal{B}}, d)$;
   (c) μ is uniformly continuous on $(\widetilde{\mathcal{B}}, d)$.

6. Prove that a subset F of L_1 is uniformly integrable if and only if the corresponding set of measures $\mathcal{F} = \{f \, dm : f \in F\}$ is equicontinuous at $\emptyset$ in $(\widetilde{\mathcal{B}}, d)$.

Chapter 14

The Dunford–Pettis Property

A Banach space X is said to have the *Dunford–Pettis property* if, whenever $x_n \xrightarrow{w} 0$ in X and $f_n \xrightarrow{w} 0$ in X^*, we have that $f_n(x_n) \to 0$. Our main result in this chapter will show that the Dunford–Pettis property is intimately related to the behavior of weakly compact operators on X. Recall that a bounded linear operator $T : X \to Y$ is said to be *weakly compact* if T maps bounded sets in X to relatively weakly compact sets in Y. Thus, T is weakly compact if and only if $\overline{T(B_X)}$ is weakly compact in Y. Since weakly compact sets are norm bounded, it follows that a weakly compact operator is bounded.

Our first result provides several equivalent characterizations of weak compactness for operators.

Theorem 14.1. *Let* $T : X \to Y$ *be bounded and linear. Then,* T *is weakly compact if and only if any one of the following hold:*

(a) $T^{**}(X^{**}) \subset Y$;

(b) $T^* : Y^* \to X^*$ *is weak*-to-weak continuous;*

(c) T^* *is weakly compact.*

Proof. Suppose that T is weakly compact. Then $T(B_X)$ is relatively weakly compact in Y. Regarding $T(B_X)$ as a subset of Y^{**}, we have

$$\overline{T(B_X)}^{(Y^{**},\text{weak}^*)} = \overline{T(B_X)}^{(Y,\text{weak})}$$

because weakly compact sets in Y are weak* compact in Y^{**}. Since $T(B_X)$ is convex, this simplifies to read

$$\overline{T(B_X)}^{\text{weak}^*} = \overline{T(B_X)} \subset Y.$$

But B_X is weak* dense in $B_{X^{**}}$ by Goldstine's theorem and T^{**} is weak*-to-weak* continuous, so

$$T^{**}(B_{X^{**}}) \subset \overline{T(B_X)}^{\text{weak}^*} \subset Y.$$

Thus, $T^{**}(X^{**}) \subset Y$.

142

Conversely, if $T^{**}(X^{**}) \subset Y$, then $T^{**}(B_{X^{**}})$ is a weakly compact subset of Y by the Banach–Alaoglu theorem and the weak*-to-weak* continuity of T^{**} (and, again, the observation that the weak* topology on Y, when considered as a subset of Y^{**}, reduces to the weak topology on Y). Thus, $T(B_X)$ is relatively weakly compact in Y, being a subset of $T^{**}(B_{X^{**}})$. This proves that T is weakly compact if and only if (a) holds.

Now, from (a),

T is weakly compact

$\Longleftrightarrow$ $T^{**}(X^{**}) \subset Y$

$\Longleftrightarrow$ $T^{**}x^{**} \in Y$ for each $x^{**} \in X^{**}$

$\Longleftrightarrow$ $T^{**}x^{**}$ is weak* continuous for each $x^{**} \in X^{**}$

$\Longleftrightarrow$ $T^{**}x^{**}(y_\alpha^*) \to T^{**}x^{**}(y^*)$, for each $x^{**} \in X^{**}$,

whenever $y_\alpha^* \xrightarrow{w^*} y^*$ in Y^*

$\Longleftrightarrow$ $x^{**}(T^*y_\alpha^*) \to x^{**}(T^*y^*)$, for each $x^{**} \in X^{**}$,

whenever $y_\alpha^* \xrightarrow{w^*} y^*$ in Y^*

$\Longleftrightarrow$ $T^*y_\alpha^* \xrightarrow{w} T^*y^*$ whenever $y_\alpha^* \xrightarrow{w^*} y^*$ in Y^*

$\Longleftrightarrow$ T^* is weak*-to-weak continuous.

This proves that T is weakly compact if and only if (b) holds.

Finally, from (b), if T is weakly compact, then T^* is weak*-to-weak continuous. But B_{Y^*} is weak* compact in Y^*, and so we have that $T^*(B_{Y^*})$ is weakly compact in X^*. Thus, T^* is weakly compact. Conversely, if T^* is weakly compact, then $T^{**}(B_{X^{**}})$ is weakly compact in Y^{**}. Thus, $T(B_X)$ is relatively weakly compact in Y. $\square$

Armed with these tools, we can now provide our main result in this chapter.

Theorem 14.2. *A Banach space X has the Dunford–Pettis property if and only if every weakly compact operator from X into a Banach space Y maps weakly compact sets in X to norm compact sets in Y; that is, if and only if every weakly compact operator is completely continuous.*

Proof. Suppose first that X has the Dunford–Pettis property and let $T : X \to Y$ be weakly compact. To begin, we consider the action of T on a weakly null sequence (x_n) in X. For each n, choose a norm one functional $g_n \in Y^*$ such that $g_n(Tx_n) = \|Tx_n\|$. That is, $T^*g_n(x_n) = \|Tx_n\|$. Now T^* is weakly compact, so there is an $f \in X^*$ and a subsequence (g_{n_k}) of (g_n) such that

$T^* g_{n_k} \xrightarrow{w} f$. But then

$$g_{n_k}(T x_{n_k}) = T^* g_{n_k}(x_{n_k}) = f(x_{n_k}) + (T^* g_{n_k} - f)(x_{n_k}) \to 0.$$

Indeed, $f(x_{n_k}) \to 0$ because (x_n) is weakly null and $(T^* g_{n_k} - f)(x_{n_k}) \to 0$ because X has the Dunford–Pettis property. Thus,

$$g_{n_k}(T x_{n_k}) = \| T x_{n_k} \| \to 0.$$

That is, $(T x_n)$ has a norm null subsequence whenever (x_n) is weakly null. By linearity, it follows that $(T x_n)$ has a norm convergent subsequence whenever (x_n) is weakly convergent. Consequently, $T(K)$ is norm compact whenever K is weakly compact.

Now suppose that every weakly compact operator on X maps weakly compact sets to norm compact sets. Given $x_n \xrightarrow{w} 0$ in X and $f_n \xrightarrow{w} 0$ in X^*, consider the map $T : X \to c_0$ defined by $T x = (f_n(x))$. It's easy to see that $T^* : \ell_1 \to X^*$ satisfies $T^* e_n = f_n$. In particular, given $x^{**} \in X^{**}$, we have

$$T^{**} x^{**}(e_n) = x^{**}(T^* e_n) = x^{**}(f_n) \to 0$$

since (f_n) is weakly null. That is, $T^{**} x^{**} \in c_0$. In other words, T is weakly compact. But then T maps weakly compact sets in X to norm compact sets in c_0, and it follows that we must have $\| T x_n \|_\infty \to 0$. (Why?) Consequently, $f_n(x_n) \to 0$. $\square$

Corollary 14.3. *Suppose that X has the Dunford–Pettis property. If $T : X \to X$ is weakly compact, then $T^2 : X \to X$ is compact.*

Proof. Since T is weakly compact, $T(B_X)$ is relatively weakly compact. And, since X has the Dunford–Pettis property, $T^2(B_X)$ is then relatively norm compact. $\square$

Corollary 14.4. *Suppose that X has the Dunford–Pettis property. If Y is a complemented reflexive subspace of X, then Y is finite dimensional.*

Proof. If $P : X \to X$ is any projection onto Y, then P is weakly compact. Indeed, $P(B_X) = B_Y$ is weakly compact since Y is reflexive. But then $P^2 = P$ is compact. Consequently, B_Y is norm compact and it follows that Y must be finite dimensional. $\square$

Corollary 14.5. *Infinite-dimensional reflexive Banach spaces do not have the Dunford–Pettis property.*

Our task in the remainder of this chapter will be to show that $C(K)$ and L_1 have the Dunford–Pettis property. Given this, it will follow that $C(K)$ and L_1 have no infinite-dimensional, reflexive, complemented subspaces.

Theorem 14.6. *Let K be a compact Hausdorff space. Then $C(K)$ has the Dunford–Pettis property.*

Proof. Suppose that $f_n \xrightarrow{w} 0$ in $C(K)$ and that $\mu_n \xrightarrow{w} 0$ in $C(K)^*$. Let

$$\lambda = \sum_{n=1}^{\infty} \frac{|\mu_n|}{2^n \|\mu_n\|},$$

where $\|\mu_n\| = |\mu_n|(K)$. Then (μ_n) is uniformly absolutely continuous with respect to λ (by Propositions 13.1 and 13.4) Thus, given $\varepsilon > 0$, there exists a $\delta > 0$ such that $|\mu_n|(A) < \varepsilon$, for all n, provided that $\lambda(A) < \delta$.

Now $f_n \to 0$ pointwise on K. Thus, by Egorov's theorem, $f_n \to 0$ uniformly on some set $K \setminus A$, where $\lambda(A) < \delta$. Next,

$$\mu_n(f_n) = \int_K f_n \, d\mu_n = \int_{K \setminus A} f_n \, d\mu_n + \int_A f_n \, d\mu_n$$

and $\int_{K \setminus A} f_n \, d\mu_n \to 0$ as $n \to \infty$ because $f_n \to 0$ uniformly on $K \setminus A$. Finally,

$$\left| \int_A f_n \, d\mu_n \right| \le \|f_n\|_\infty \cdot |\mu_n|(A) \le \varepsilon \cdot \sup_k \|f_k\|_\infty$$

because $\lambda(A) < \delta$. Thus, $\mu_n(f_n) \to 0$. $\square$

Corollary 14.7. *If X is a separable, infinite-dimensional, reflexive Banach space, then X is isometric to an uncomplemented subspace of $C[0, 1]$.*

We next attack the Dunford–Pettis property in L_1. To this end, we will need some additional information about weak convergence in L_∞.

Proposition 14.8. *Suppose that $g_n \xrightarrow{w} 0$ in L_∞. Then, given $\varepsilon > 0$, there exists a Borel set $A \subset [0, 1]$ with $m(A) > 1 - \varepsilon$ such that $g_n \to 0$ uniformly on A.*

Proof. By repeated application of Lusin's theorem, we can find a Borel subset B of $[0, 1]$, with $m(B) > 1 - \varepsilon/2$, and a sequence of functions $\tilde{g}_n$, each continuous on B, such that $g_n = \tilde{g}_n$ a.e. on B. Moreover, by the Lebesgue density theorem, we may assume that each point $x \in B$ has density 1; that is,

$$\lim_{r \to 0} \frac{m((x - r, x + r) \cap B)}{2r} = 1$$

for each $x \in B$ (thus, in particular, B has no isolated points). It follows that

$$\left| \sum_{k=1}^{n} a_k \tilde{g}_k(x) \right| \leq \underset{B}{\text{ess.sup}} \left| \sum_{k=1}^{n} a_k g_k \right| \leq \left\| \sum_{k=1}^{n} a_k g_k \right\|_{\infty}$$

for all scalars (a_k) and all $x \in B$. Hence, for each $x \in B$, the map $g_n \mapsto \tilde{g}_n(x)$ extends to a bounded linear functional on $[g_n]$. But then, since (g_n) is weakly null, we must have $\tilde{g}_n(x) \to 0$ as $n \to \infty$ for each $x \in B$. Hence, $g_n \to 0$ a.e. on B. Finally, by Egorov's theorem, there exists a Borel set $A \subset B$ with $m(A) > 1 - \varepsilon$ such that $g_n \to 0$ uniformly on A. $\square$

Corollary 14.9. L_1 *has the Dunford–Pettis property.*

Proof. Let $f_n \xrightarrow{w} 0$ in L_1 and let $g_n \xrightarrow{w} 0$ in L_∞. Then (f_n) is uniformly integrable in L_1. Thus, given $\varepsilon > 0$, there is a $\delta > 0$ such that $\int_B |f_n| < \varepsilon$, for all n provided that $m(B) < \delta$. By Proposition 14.8, there is a Borel set A with $m(A) > 1 - \delta$ such that $g_n \to 0$ uniformly on A. Thus,

$$\left| \int_0^1 f_n g_n \right| \leq \int_{A^c} |f_n g_n| + \int_A |f_n g_n|$$
$$\leq \varepsilon \cdot \sup_n \|g_n\|_\infty + \sup_n \|g_n \chi_A\|_\infty \cdot \sup_n \|f_n\|_1,$$

which tends to 0 as $n \to \infty$. $\square$

Corollary 14.10. *Every complemented, infinite-dimensional subspace of L_1 is nonreflexive.*

Notes and Remarks

Our presentation in this chapter borrows heavily from some unpublished notes for a course on Banach space theory given by Stephen Dilworth at the University of South Carolina [35]. Theorem 14.1 is often called Gantmacher's theorem, after Vera Gantmacher, who proved the theorem for separable spaces [52]; the general case was later settled by Nakamura [107]. Theorem 14.9 (in different language) is due to Dunford and Pettis [41]. The Dunford–Pettis property was so named by Grothendieck [62], who gave us Theorem 14.2, Theorem 14.6, and a wealth of other results. For more on the Dunford–Pettis property (as well as its history), see Diestel and Uhl [34].

Exercises

1. If $T : X \to Y$ is bounded and linear, prove that $T^{**} : X^{**} \to Y^{**}$ is weak*-to-weak* continuous.

2. If K is weakly compact in X, prove that K is weak* closed as a subset of X^{**}.

3. If K is a subset of X such that the weak* closure of K is again contained in X (when considered as a subset of X^{**}), prove that K is relatively weakly compact.

4. If $K \subset X$ is weak* compact as a subset of X^{**}, prove that K is weakly compact in X.

5. Prove that ℓ_1 has the Dunford–Pettis property.

6. Let A be a Borel subset of $[0, 1]$ with Lebesgue density 1 and let $g : A \to \mathbb{R}$ be continuous. Prove that $|g(x)| \le \text{ess.sup}_1 g|$ for every $x \in A$.

7. Prove that if X^* has the Dunford–Pettis property, then X does too. Thus, c_0 has the Dunford–Pettis property.

8. Prove that a Banach space X has the Dunford–Pettis property if and only if every weakly compact operator $T : X \to c_0$ maps weakly compact sets in X to norm compact sets in c_0. (In other words, we need only consider $Y = c_0$ in Theorem 14.2.)

Chapter 15
$C(K)$ Spaces II

By now, even a skeptical reader should be thoroughly sold on the utility of embeddings into cubes. But the sales pitch is far from over! We next pursue the consequences of a result stated earlier: If X is completely regular, then $C_b(X)$ completely determines the topology on X. In brief, to know $C_b(X)$ is to know X. Just how far can this idea be pushed? If $C_b(X)$ and $C_b(Y)$ are isomorphic (as Banach spaces, as lattices, or as rings), must X and Y be homeomorphic? Which topological properties of X can be attributed to structural properties of $C_b(X)$ (and conversely)?

These questions were the starting place for Marshall Stone's 1937 landmark paper, "Applications of the Theory of Boolean Rings to General Topology" [140]. It's in this paper that Stone gave his account of the Stone–Weierstrass theorem, the Banach–Stone theorem, and the Stone–Čech compactification. (These few are actually tough to find among the dozens of results in this mammoth 106-page work.) A signal passage from his introduction may be paraphrased as follows: "We obtain a reasonably complete algebraic insight into the structure of $C_b(X)$ and its correlation with the structure of the underlying topological space." Stone's work proved to be a gold mine – the digging continued for years! – and its influence on algebra, analysis, and topology alike can be seen in virtually every modern textbook.

Independently, but later that same year (1937), Eduard Čech [24] gave another proof of the existence of the compactification but, strangely, credits a 1929 paper of Tychonoff for the result (see Shields [136] for more on this story). To a large extent, we will be faithful to Čech's approach, which leans more toward topology than algebra.

The Stone–Čech Compactification

Given a Hausdorff topological space X, any compact Hausdorff space Y that contains a dense subspace homeomorphic to X is called a *Hausdorff compactification* for X. What this means, in practice, is that we look for any compact

Hausdorff space Y that admits a homeomorphic embedding $f : X \to Y$ from X into Y; the closure of $f(X)$ in Y then defines a compactification of X.

There is a hierarchy of compactifications, the full details of which aren't necessary just now. It shouldn't come as a surprise that, for locally compact spaces, the one-point compactification is the smallest compactification in this hierarchy. What we're after is the largest compactification.

Given a completely regular space X, we define βX, *the Stone–Čech compactification of* X, to be the *closure* of $e(X)$ in $[0, 1]^{C(X;[0,1])}$. From the Embedding lemma (Lemma 12.3), X is then homeomorphic to a dense subset of the compact Hausdorff space βX. Note that if X is compact, then e is a homeomorphism from X *onto* βX.

Strictly speaking, the compactification is defined to be the pair $(e, \beta X)$, but we will have little need for such formality. In fact, we often just think of X as already living inside βX and simply ignore the embedding e.

βX is characterized by the following extension theorem:

Theorem 15.1 (Extension Theorem). *Let X be a completely regular space and let $e : X \to \beta X$ be the canonical embedding.*

(a) *Every bounded, continuous function $f : X \to \mathbb{R}$ extends to a continuous function $F : \beta X \to \mathbb{R}$ in the sense that $F \circ e = f$.*

(b) *If Y is a compact Hausdorff space, then each continuous function $f : X \to Y$ extends to a continuous function $F : \beta X \to Y$ in the sense that $F \circ e = f$. If Y is a compactification of X, then F is onto. In particular, every Hausdorff compactification of X is the continuous image of βX.*

Proof. (a): Suppose that $f : X \to \mathbb{R}$ is continuous and bounded. By composing with a suitable homeomorphism of $\mathbb{R}$, we may assume that $f : X \to [0, 1]$. But then f is one of the "coordinates" in the product space $[0, 1]^{C(X;[0,1])}$. We claim that the coordinate projection π_f, when restricted to βX, is the extension we want. Indeed, $\pi_f : \beta X \to [0, 1]$ is continuous, and π_f, when restricted to $e(X)$, is just f since $\pi_f(e(x)) = e(x)(f) = f(x)$. Note that $F = \pi_f|_{\beta X}$ is unique because $e(X)$ is dense in βX.

(b): Now suppose that Y is a compact Hausdorff space and that $f : X \to Y$ is continuous. Let $C_X = C(X; [0, 1])$ and $C_Y = C(Y; [0, 1])$, and let $e_X : X \to [0, 1]^{C_X}$ and $e_Y : Y \to [0, 1]^{C_Y}$ be the canonical embeddings (which we may consider as maps into βX and βY, respectively). Note that since Y is compact, e_Y is a homeomorphism from Y onto βY.

In order to extend f to βX, we first "lift" f to a mapping Φ from $[0, 1]^{C_X}$ to $[0, 1]^{C_Y}$. To better understand this lifting, recall that $f(x) \in Y$ corresponds to the "tuple" $(g(f(x))_{g \in C_Y}$ in $[0, 1]^{C_Y}$ under the evaluation map e_Y. Also note that $g \circ f \in C_X$ whenever $g \in C_Y$.

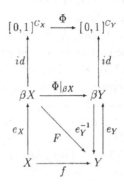

We now define $\Phi : [0, 1]^{C_X} \to [0, 1]^{C_Y}$ by specifying the coordinates of its images: $\pi_g(\Phi(p)) = \pi_{g \circ f}(p)$ for each $g \in C_Y$ and each $p \in [0, 1]^{C_X}$. The map Φ is continuous (since its coordinates are) and $\Phi \circ e_X = e_Y \circ f$:

$$\pi_g(\Phi(e_X(x))) = \pi_{g \circ f}(e_X(x)) = g(f(x)) = \pi_g(e_Y(f(x))).$$

Next, because $\Phi(e_X(x)) = e_Y(f(x)) \in e_Y(Y)$, we have that $\Phi(\beta X) \subset \overline{e_Y(Y)} = e_Y(Y)$. Consequently, $F = e_Y^{-1} \circ \Phi|_{\beta X} \circ e_X$ extends f. Again, uniqueness of F follows from the fact that $e(X)$ is dense in βX.

Finally, if $f : X \to Y$ is a homeomorphism from X onto a dense subspace of Y, then $F : \beta X \to Y$ maps βX onto a dense, compact subset of Y. That is, F is onto. $\square$

The conclusion of part (b) of the Extension theorem is that the Stone–Čech compactification is the largest Hausdorff compactification of X (in a categorical sense).

The Extension theorem tells us something new about the space $C_b(X)$; note that the map $F \mapsto F \circ e$ defines a linear isometry from $C(\beta X)$ *onto* $C_b(X)$. That is, each $f \in C_b(X)$ is of the form $F \circ e$ for some $F \in C(\beta X)$. In particular, we now know that $C_b(X)$ is a $C(K)$ space for some (rather specific!) compact Hausdorff space K.

Corollary 15.2. *Let X be completely regular.*

(i) *$C_b(X)$ is isometrically isomorphic to $C(\beta X)$.*

(ii) *If Y is completely regular, then each continuous function $f : X \to Y$*

lifts to a continuous function $F : \beta X \to \beta Y$ *satisfying* $F \circ e_X = e_Y \circ f$.

(iii) *If Y is any Hausdorff compactification of X enjoying the property of βX described in part* (a) *of the Extension theorem, then Y is homeomorphic to βX.*

Proof. Only (iii) requires a proof. Let Y be a compact Hausdorff space, and suppose that $f : X \to Y$ is a homeomorphism from X onto a dense subspace of Y. Suppose further that each $h \in C_b(X)$ extends to a continuous function $g \in C(Y)$ with $g \circ f = h$. Then, in particular, each $h \in C(X; [0, 1])$ is of the form $g \circ f$ for some $g \in C(Y; [0, 1])$. This implies that the "lifting" Φ of f constructed in the proof of part (b) of the Extension theorem is one-to-one. Thus, the extension $F : \beta X \to Y$ of f is both one-to-one and onto and hence is a homeomorphism. $\square$

For later reference, we next present two simple methods for computing the Stone–Čech compactification.

Lemma 15.3. *Let X be completely regular.*

(a) *Let $T \subset X \subset \beta X$. If each bounded, continuous, real-valued function on T extends continuously to X, then $\beta T = \mathrm{cl}_{\beta X} T$ (the closure of T in βX).*

(b) *If $X \subset T \subset \beta X$, then $\beta T = \beta X$.*

Proof. (a): By design, each bounded, continuous, real-valued function on T extends all the way to βX and hence also to $\mathrm{cl}_{\beta X} T$. Since $\mathrm{cl}_{\beta X} T$ is a compactification of T, it must, then, be βT by Corollary 15.2 (iii).

(b): First observe that T is dense in βX, and so βX is a compactification of T. Next, each bounded, continuous, real-valued function on T extends continuously to βX since its restriction to X does. The fact that X is dense in T takes care of any uniqueness problems caused by extending the restriction (or restricting the extension . . .). Thus, by Corollary 15.2 (iii), $\beta X = \beta T$. $\square$

We're way overdue for a few concrete examples.

Examples

1. $\beta(0, 1) \neq [0, 1]$. Why? Because $\sin(1/x)$ has no continuous extension to $[0, 1]$! But $\sin(1/x)$ does, of course, extend continuously to $\beta(0, 1)$, whatever that is! As we'll see in a moment, $\beta(0, 1)$ is much larger than $[0, 1]$.

2. If D is any discrete space, then $\ell_\infty(D) = C_b(D) = C(\beta D)$ isometrically. In particular, $\ell_\infty = C(\beta\mathbb{N})$. Since ℓ_∞ isn't separable, we now have a proof that $\beta\mathbb{N}$ isn't metrizable. In fact, as we'll see, $\beta\mathbb{N}$ is in no way sequentially compact.

3. $\text{card}(\beta\mathbb{N}) = 2^c$. Here's a clever proof: Recall that $Y = [0, 1]^{[0,1]}$ is separable. Hence, there is a *continuous* map from $\mathbb{N}$ onto a dense subset of Y. This map extends to a continuous map from $\beta\mathbb{N}$ onto all of Y. Consequently, $2^c = \text{card}(Y) \leq \text{card}(\beta\mathbb{N})$, while from the construction of $\beta\mathbb{N}$ it follows that $\text{card}(\beta\mathbb{N}) \leq \text{card}([0, 1]^{[0,1]^{\mathbb{N}}}) = 2^c$.

4. $\text{card}(\beta(0, 1)) = \text{card}(\beta\mathbb{R}) = \text{card}(\beta\mathbb{N})$. The first equality is obvious since $(0, 1)$ is homeomorphic to $\mathbb{R}$. Next, as in 3, $\beta\mathbb{R}$ is the continuous image of $\beta\mathbb{N}$ because $\mathbb{R}$ is separable. Thus, $\text{card}(\beta\mathbb{R}) \leq \text{card}(\beta\mathbb{N})$. To finish the proof, we'll find a copy of $\beta\mathbb{N}$ living inside $\beta\mathbb{R}$. Here's how: Each bounded (continuous) real-valued function on $\mathbb{N}$ extends to a bounded continuous function on all of $\mathbb{R}$. (No deep theorems needed here! Just "connect the dots.") Thus, by Lemma 15.3 (a), $\text{cl}_{\beta\mathbb{R}}\mathbb{N} = \beta\mathbb{N}$. Hence, $\text{card}(\beta\mathbb{R}) \geq \text{card}(\beta\mathbb{N})$.

5. Banach limits (invariant means). We will use the fact that $\ell_\infty = C(\beta\mathbb{N})$ to extend the notion of "limit" to include all bounded sequences. First, given $x \in \ell_\infty$, let's agree to write $\tilde{x}$ for its unique extension to an element of $C(\beta\mathbb{N})$. Now, given any (fixed) point $t \in \beta\mathbb{N} \setminus \mathbb{N}$, we define: $\text{Lim } x = \tilde{x}(t)$. This generalized limit, often called a *Banach limit*, satisfies

$$\text{Lim } x = \lim x, \text{ if } x \text{ actually converges};$$
$$\liminf x \leq \text{Lim } x \leq \limsup x;$$
$$\text{Lim}\,(ax + by) = a\,\text{Lim}\,x + b\,\text{Lim}\,y;$$
$$\text{Lim}\,(x\,y) = (\text{Lim}\,x)\,(\text{Lim}\,y).$$

Just for fun, let's check the first claim. The key here is that t is in the closure of $\{n : n \geq m\}$ for each m. Thus, if $L = \lim x$ exists and if $\varepsilon > 0$, then $x_n \in [L - \varepsilon, L + \varepsilon]$ for all $n \geq m$, for some m. Hence, $\tilde{x}(t) \in [L - \varepsilon, L + \varepsilon]$, too. That is, $\tilde{x}(t) = L$.

What we've actually found is a (particularly convenient) Hahn–Banach extension of the functional "lim" on the subspace c of ℓ_∞. What makes this example interesting, as we'll see in Chapter 16, is that no point $t \in \beta\mathbb{N} \setminus \mathbb{N}$ can be the limit of a *sequence* in $\mathbb{N}$.

6. From our discussion of completely regular spaces in the last chapter, we can give an alternate definition of the Stone–Čech compactification. Each completely regular space T lurks within $C_b(T)^*$ under the guise of the point masses: $P = \{\delta_t : t \in T\}$. It follows that we can define

βT to be the weak* closure of P in $C_b(T)^*$. Why? Well, since $\overline{P} =$ weak*-cl P is a compactification of T, we only need to show that each element $f \in C_b(T)$ extends to an element $\hat{f} \in C\left(\overline{P}\right)$, and this is easier than it might sound: Just define $\hat{f}(p)$ to be $p(f)$! In other words, the canonical embedding of $C_b(T)$ into $C_b(T)^{**}$ supplies an embedding of $C_b(T)$ into $C(\beta T)$.

7. Finally, here's a curious proof of (a special case of) Tychonoff's theorem based on the Extension theorem: Let $X = \prod_{\alpha \in A} X_\alpha$, where each X_α is a (nonempty) compact Hausdorff space. Then, X is completely regular. Hence, the projection maps $\pi_\alpha : X \to X_\alpha$ have continuous extensions $\tilde{\pi}_\alpha : \beta X \to X_\alpha$. But this means that the map $p \mapsto (\tilde{\pi}_\alpha(p))_{\alpha \in A}$, from βX *onto* X, is continuous! Thus, X is compact.

Return to $C(K)$

$\beta \mathbb{N}$ is a most curious space and will play a major role in the next chapter. More generally, as our examples might suggest, the Stone–Čech compactification of a discrete space is a potentially useful tool. This is further highlighted by the following observation:

Lemma 15.4. *Every compact Hausdorff space K is the continuous image of βD for some discrete space D. Consequently, $C(K)$ is isometric to a subspace of $C(\beta D) = \ell_\infty(D)$.*

Proof. Let D_0 be any dense set in K, and let D be D_0 with the discrete topology. Then, the formal identity from D into K extends to a continuous map φ from βD *onto* K! The composition map $f \mapsto f \circ \varphi$ defines a linear isometry from $C(K)$ into $C(\beta D)$. $\square$

As an immediate corollary, we get a result that we've (essentially) seen before.

Corollary 15.5. *If K is a compact metric space, then $C(K)$ embeds isometrically into ℓ_∞. Hence, every separable normed linear space embeds isometrically into ℓ_∞.*

In the next chapter we will compute the dual of $C(K)$ by, instead, computing the dual of $\ell_\infty(D)$. That is, we will prove the Riesz representation theorem for ℓ_∞ spaces and then transfer our work to the $C(K)$ spaces. This being the case, we might be wise to quickly summarize a few features of the "fancy" versions of the Riesz representation theorem.

First, the space $C(K)$ and its relatives $C_C(X)$, $C_b(T)$, and so on, are probably best viewed as *vector lattices*. Under the usual pointwise ordering of functions, $C(K)$ is an ordered vector space:

$$f, g \in C(K), \ f \geq g \Longrightarrow f + h \geq g + h \text{ for all } h \in C(K), \text{ and}$$
$$f \in C(K), \ a \in \mathbb{R}, \ f \geq 0, \ a \geq 0 \Longrightarrow af \geq 0.$$

In addition, $C(K)$ is a lattice:

$$f, g \in C(K) \Longrightarrow f \vee g = \max\{f, g\} \text{ and } f \wedge g = \min\{f, g\} \text{ are in } C(K).$$

This last observation allows us to define positive and negative parts: $f^+ = f \vee 0$ and $f^- = -(f \wedge 0)$. Thus, each $f \in C(K)$ can be written as $f = f^+ - f^-$. Moreover, $|f| = f^+ + f^-$.

Now $C(K)$ enjoys the additional property that its usual norm is compatible with the order structure in the sense that

$$|f| \leq |g| \Longrightarrow \|f\|_\infty \leq \|g\|_\infty.$$

Since $C(K)$ is also complete under this norm, we say that $C(K)$ is a *Banach lattice*.

As it happens, the dual of a Banach lattice can be given an order structure too: We define an order on $C(K)^*$ by defining $S \geq T$ to mean that $S(f) \geq T(f)$ for all $f \geq 0$. In particular, a linear functional T on $C(K)$ is *positive* if $T(f) \geq 0$ whenever $f \geq 0$. It's easy to see that every positive linear functional is bounded; indeed, if T is positive, then

$$|T(f)| \leq T(|f|) \leq T(\|f\|_\infty \cdot 1) = \|f\|_\infty \cdot T(1),$$

where 1 denotes the constant 1 function. Hence, $\|T\| = T(1)$.

What's a little harder to see is that the dual space will again be a Banach lattice under this order. We first need to check that each bounded linear functional can be written as the difference of positive functionals. Given $T \in C(K)^*$, we define

$$T^+(f) = \sup\{T(g) : 0 \leq g \leq f\} \text{ for } f \geq 0.$$

It's tedious, but not difficult, to check that T^+ is additive on positive elements and that $T^+(af) = aT^+(f)$ for $a \geq 0$. Now for arbitrary f we define $T^+(f) = T^+(f^+) - T^+(f^-)$. It follows that T^+ is positive and linear and satisfies $T^+(f) \geq T(f)$ for every $f \geq 0$. Thus, $T^- = T^+ - T$ is likewise positive and linear. That is, we've written T as the difference of positive linear functionals. Consequently, *a linear functional on $C(K)$ is bounded if and only if it can be written as the difference of positive linear functionals.*

Finally, let's compute the norm of T in terms of T^+ and T^-. Clearly,

$$\|T\| \le \|T^+\| + \|T^-\| = T^+(1) + T^-(1).$$

On the other hand, given $0 \le f \le 1$, we have $|2f - 1| \le 1$ and hence $\|T\| \ge T(2f - 1) = 2T(f) - T(1)$. By taking the supremum over all $0 \le f \le 1$ we get

$$\|T\| \ge 2T^+(1) - T(1) = T^+(1) + T^-(1).$$

Hence, $\|T\| = T^+(1) + T^-(1)$. If we define $|T| = T^+ + T^-$, as one would expect, then we have $\|T\| = \||T|\| = |T|(1)$. It follows from this definition of $|T|$ that $C(K)^*$ is itself a Banach lattice.

In terms of the Riesz representation theorem, all of this tells us that we only need to represent the positive linear functionals on $C(K)$. As you no doubt already know, each positive linear functional on $C(K)$ will turn out to be integration against a positive measure. The generic linear functional will then be given by integration against a signed measure. In terms of measures, $\mu = \mu^+ - \mu^-$ is the Jordan decomposition of μ, whereas $|\mu| = \mu^+ + \mu^-$ is the total variation of μ. Not surprisingly, we define $\|\mu\| = |\mu|(K) = \int_K 1\, d|\mu| = \||\mu|\|$.

Notes and Remarks

The Stone–Čech compactification is discussed in any number of books; see, for example, Folland [48, Chapter 4], Gillman and Jerison [55], Wilansky [145], or Willard [146]. For more on Banach limits and their relationship to $\beta\mathbb{N}$, see Nakamura and Kakutani [108]. Banach lattices are treated in several books; see, for example, Aliprantis and Burkinshaw [3], Lacey [88], Meyer-Nieberg [102], or Schaefer [130].

Exercises

1. Prove Corollary 15.2 (ii).

2. Let X be completely regular. Show that X is locally compact if and only if X is open in βX.

3. Complete the proof of the claims made in Example 5 concerning the Banach limit $\mathrm{Lim}\, x$ on ℓ_∞.

Chapter 16
$C(K)$ Spaces III

In this chapter we present Garling's proof [53] of the *Riesz representation theorem* for the dual of $C(K)$, K compact Hausdorff. This theorem goes by a variety of names: The Riesz–Markov theorem, the Riesz–Kakutani theorem, and others. The version that we'll prove states:

Theorem 16.1. *Let K be a compact Hausdorff space, and let T be a positive linear functional on $C(K)$. Then there exists a unique positive Baire measure μ on K such that $T(f) = \int_K f \, d\mu$ for every $f \in C(K)$.*

As we pointed out in the last chapter, our approach will be to first prove the theorem for ℓ_∞ spaces. To this end, we will need to know a bit more about the Stone–Čech compactification of a discrete space and a bit more measure theory. First the topology.

The Stone–Čech Compactification of a Discrete Space

A topological space is said to be *extremally disconnected*, or *Stonean*, if the closure of every open set is again open. Obviously, discrete spaces are extremally disconnected. Less mundane examples can be manufactured from this starting point:

Lemma 16.2. *If D is a discrete space, then βD is extremally disconnected.*

Proof. Let U be open in βD, and let $A = U \cap D$. Then A is dense in U since U is open, and so $\mathrm{cl}_{\beta D} A = \mathrm{cl}_{\beta D} U$. Now we just check that $\mathrm{cl}_{\beta D} A$ is also open. The characteristic function $\chi_A : D \to \{0, 1\}$ (a continuous function on D!) extends continuously to some $f : \beta D \to \{0, 1\}$. Thus, by continuity, $\mathrm{cl}_{\beta D} A = f^{-1}(\{1\})$ is open. $\square$

156

By modifying this proof, it's not hard to show that a completely regular space X is extremally disconnected if and only if βX is extremally disconnected. We leave the proof of this claim as an exercise.

Notice that if A and B are disjoint (open) sets in a discrete space D, then $\mathrm{cl}_{\beta D} A$ and $\mathrm{cl}_{\beta D} B$ are disjoint in βD. Indeed, just as in the proof of Lemma 16.2, the function χ_A extends continuously to a function $f : \beta D \to \{0, 1\}$, which satisfies $\mathrm{cl}_{\beta D} A = f^{-1}(\{1\})$ and $\mathrm{cl}_{\beta D} B \subset f^{-1}(\{0\})$. In particular, any set of the form $\mathrm{cl}_{\beta D} A$, where $A \subset D$, is *clopen*; that is, simultaneously open and closed. In fact, every clopen subset of βD is of this same form.

Lemma 16.3. *Let D be a discrete space. Then the clopen subsets of βD are of the form $\mathrm{cl}_{\beta D} A$, where A is open in D. Further, the clopen sets form a base for the topology of βD.*

Proof. If C is a clopen subset of βD, then, just as in Lemma 16.2,

$$C = \mathrm{cl}_{\beta D} C = \mathrm{cl}_{\beta D}(C \cap D).$$

Now let U be an open set in βD, and let $x \in U$. Since βD is regular, we can find a neighborhood V of x such that $x \in V \subset \mathrm{cl}_{\beta D} V \subset U$. Because $\mathrm{cl}_{\beta D} V$ is clopen, this finishes the proof. $\square$

A Few Facts About $\beta\mathbb{N}$

We can now shed a bit more light on $\beta\mathbb{N}$. Note, for example, that $\mathbb{N}$ is *open* in $\beta\mathbb{N}$. Indeed, given $n \in \mathbb{N}$, the set $\mathrm{cl}_{\beta\mathbb{N}}\{n\}$ is open in $\beta\mathbb{N}$. But $\{n\}$ is compact; hence, $\{n\} = \mathrm{cl}_{\beta\mathbb{N}}\{n\}$. That is, $\{n\}$ is open in $\beta\mathbb{N}$ too. In particular, each $n \in \mathbb{N}$ is an isolated point in $\beta\mathbb{N}$.

It follows that a sequence in $\mathbb{N}$ converges in $\beta\mathbb{N}$ if and only if it is *eventually constant*; that is, if and only if it already converges in $\mathbb{N}$. Suppose, to the contrary, that (x_n) is a sequence in $\mathbb{N}$ that is not eventually constant, and suppose that (x_n) converges to a point $t \in \beta\mathbb{N} \backslash \mathbb{N}$. Then the range of (x_n) must be infinite, for otherwise (x_n) would have a subsequence converging in $\mathbb{N}$. Thus, by induction, we can choose a subsequence (x_{n_k}) of distinct integers; $x_{n_i} \neq x_{n_j}$ if $i \neq j$. But now the sets $A = \{x_{n_{2k}}\}$ and $B = \{x_{n_{2k-1}}\}$ are disjoint in $\mathbb{N}$, while t is in the closure of each, which is a contradiction. In particular, we've shown that no point $t \in \beta\mathbb{N} \backslash \mathbb{N}$ can be the limit of a sequence in $\mathbb{N}$. As a consequence, $\beta\mathbb{N}$ isn't sequentially compact (and thus isn't metrizable).

A similar argument shows that, for any $t \in \beta\mathbb{N} \backslash \mathbb{N}$, the compact set $\{t\}$ isn't a G_δ in $\beta\mathbb{N}$. Indeed, if $\{t\}$ were a G_δ, then we could find a sequence of clopen sets of the form $B_n = \mathrm{cl}_{\beta\mathbb{N}} A_n$, where $A_n \subset \mathbb{N}$, such that $\{t\} = \bigcap_{n=1}^{\infty} B_n$. The sets (A_n) have the finite intersection property, and so we can choose a sequence of distinct points $x_n \in A_1 \cap \cdots \cap A_n$. Putting $A = \{x_n : n \in \mathbb{N}\}$, we would then have $\mathrm{cl}_{\beta\mathbb{N}} A \backslash \mathbb{N} \subset \bigcap_{n=1}^{\infty} B_n = \{t\}$. But since A is an infinite subset of $\mathbb{N}$, it follows that $\mathrm{cl}_{\beta\mathbb{N}} A$ is homeomorphic to $\beta\mathbb{N}$ and, in particular, has cardinality 2^c, which is a contradiction. What we've shown, of course, is that every nonempty G_δ subset of $\beta\mathbb{N} \backslash \mathbb{N}$ has cardinality 2^c. This observation will be of interest in our discussion of measures on $\beta\mathbb{N}$.

If A is an infinite subset of $\mathbb{N}$, then $\mathrm{cl}_{\beta\mathbb{N}} A$ is homeomorphic to $\beta\mathbb{N}$. Since $\mathbb{N}$ can be partitioned into infinitely many disjoint, infinite subsets (A_n), it follows that $\beta\mathbb{N}$ contains infinitely many pairwise disjoint clopen sets $B_n = \mathrm{cl}_{\beta\mathbb{N}} A_n$, each homeomorphic to $\beta\mathbb{N}$. Note, however, that $T = \bigcup_{n=1}^{\infty} B_n$ is not all of $\beta\mathbb{N}$ because a compact space can't be written as a disjoint union of infinitely many disjoint open sets. Since $\mathbb{N} \subset T \subset \beta\mathbb{N}$, we do have that T is dense in $\beta\mathbb{N}$; moreover, $\beta T = \beta\mathbb{N}$.

Using this observation, we can build a copy of $\beta\mathbb{N}$ inside the closed set $\beta\mathbb{N} \backslash \mathbb{N}$. To see this, let $t_n \in B_n \backslash \mathbb{N} = \mathrm{cl}_{\beta\mathbb{N}} A_n \backslash \mathbb{N}$ and set $D = \{t_n : n \geq 1\}$. Obviously, D is a discrete subspace of $\beta\mathbb{N}$ and, as such, is homeomorphic to $\mathbb{N}$. Thus, βD is homeomorphic to $\beta\mathbb{N}$. Now, given a bounded, real-valued function f on D, we can easily extend f to a bounded continuous function on T by setting $f(x) = f(t_n)$ for every $x \in B_n$. Since $D \subset T \subset \beta\mathbb{N} = \beta T$, it follows that $\beta D = \mathrm{cl}_{\beta\mathbb{N}} D$. But since D is a subset of the closed set $\beta\mathbb{N} \backslash \mathbb{N}$, so is βD. In short, we've just found a copy of $\beta\mathbb{N}$ in $\beta\mathbb{N} \backslash \mathbb{N}$.

As a very clever argument demonstrates, there are, in fact, $\mathfrak{c}$ disjoint copies of $\beta\mathbb{N}$ living inside $\beta\mathbb{N} \backslash \mathbb{N}$. Indeed, recall that we can find $\mathfrak{c}$ subsets $(E_\alpha)_{\alpha \in A}$ of $\mathbb{N}$ such that each E_α is infinite, and any two E_α have, at most, a *finite* intersection. For each α, the set $F_\alpha = \mathrm{cl}_{\beta\mathbb{N}} E_\alpha \backslash \mathbb{N}$ is then homeomorphic to $\beta\mathbb{N} \backslash \mathbb{N}$ and so contains a copy of $\beta\mathbb{N}$. Finally, notice that the F_α are pairwise disjoint because, for each $\alpha \neq \gamma$, the set $\mathrm{cl}_{\beta\mathbb{N}}(E_\alpha \backslash E_\gamma)$ differs from $\mathrm{cl}_{\beta\mathbb{N}} E_\alpha$ in only a finite subset of $\mathbb{N}$.

"Topological" Measure Theory

Now for some measure theory. Our job, remember, is to compute the dual of $C(\beta D)$, where D is discrete. We know that there are enough clopen sets in βD to determine its topology completely and, so, enough clopen sets to determine $C(\beta D)$ completely. It should come as no surprise, then, that there are also enough clopen sets to determine $C(\beta D)^*$. The clopen sets in βD form an *algebra* of sets, which we will denote by $\mathcal{A}$; the σ-algebra generated by $\mathcal{A}$ will be denoted by Σ. Two more σ-algebras will enter the picture: $\mathcal{B}$, the Borel σ-algebra on βD, and $\mathcal{B}_0$, the Baire σ-algebra on βD. The Baire σ-algebra is the smallest σ-algebra $\mathcal{B}_0$ such that each $f \in C(\beta D)$ is $\mathcal{B}_0$-measurable. It's not hard to see that $\mathcal{B}_0$ and Σ are sub-σ-algebras of $\mathcal{B}$. The next lemma shows that we also have $\mathcal{B}_0 \subset \Sigma$:

Lemma 16.4. *Each $f \in C(\beta D)$ is Σ-measurable. Moreover, the simple functions based on clopen sets in $\mathcal{A}$ are uniformly dense in $C(\beta D)$.*

Proof. Let $f \in C(\beta D)$ and let $\alpha \in \mathbb{R}$. Then,

$$\{f \geq \alpha\} = \bigcap_{n=1}^{\infty} \{f > \alpha - 1/n\} \subset \bigcap_{n=1}^{\infty} \overline{\{f > \alpha - 1/n\}}$$

$$\subset \bigcap_{n=1}^{\infty} \{f \geq \alpha - 1/n\} \quad \text{(by continuity)}$$

$$= \{f \geq \alpha\}.$$

Thus, we have equality throughout. It then follows from Lemma 16.2 that the set $\{f \geq \alpha\}$ is the countable intersection of clopen sets and, as such, is in Σ. Hence, f is Σ-measurable.

The second assertion follows from the fact that the finitely-many-valued functions are dense in $\ell_\infty(D)$. $\square$

Because each $f \in C(\beta D)$ is Σ-measurable, we must have $\mathcal{B}_0 \subset \Sigma$. On the other hand, since each clopen subset of βD can be realized as a "zero set" for some $f \in C(\beta D)$, we also have $\mathcal{A} \subset \mathcal{B}_0$, and hence $\Sigma \subset \mathcal{B}_0$. Thus, the σ-algebra of Baire sets on βD coincides with the σ-algebra generated by the clopen sets in βD.

Please note that our proof also shows that $\{f \geq \alpha\}$ is a compact G_δ in βD. The Baire σ-algebra on any "reasonable" space turns out to be the σ-algebra generated by the compact G_δ sets. In contrast, note that the Borel σ-algebra on any compact Hausdorff space could be defined as the σ-algebra generated by the compact sets. We briefly describe a few such cases below.

For a locally compact space X, the Baire σ-algebra $\mathcal{B}_0$ is defined to be the smallest σ-algebra on X such that each element of $C_C(X)$ is measurable, where $C_C(X)$ is the space of continuous real-valued functions on X with compact support.

Lemma 16.5. *Let X be a locally compact Hausdorff space.*

(a) *If $f \in C_C(X)$ is nonnegative, then $\{f \geq \alpha\}$ is a compact G_δ for every $\alpha > 0$.*

(b) *If K is a compact G_δ in X, then there is an $f \in C_C(X)$ with $0 \le f \le 1$ such that $K = f^{-1}(\{1\})$.*

(c) *The Baire σ-algebra in X is the σ-algebra generated by the compact G_δ sets in X.*

Proof. (a): For $\alpha > 0$, the set $\{f \ge \alpha\}$ is a closed subset of the support of f and hence is compact. And, as before, $\{f \ge \alpha\} = \bigcap_{n=1}^{\infty}\{f > \alpha - 1/n\}$ is also a G_δ.

(b): Suppose that $K = \bigcap_{n=1}^{\infty} U_n$, where U_n is open. Apply Urysohn's lemma to find an $f_n \in C_C(X; [0, 1])$ with $f_n = 1$ on K and $f_n = 0$ off U_n. Then, $f = \sum_{n=1}^{\infty} 2^{-n} f_n$ is in $C_C(X; [0, 1])$ and $f = 1$ precisely on K.

(c): Let $\mathcal{G}$ be the σ-algebra generated by the compact G_δ sets in X. From (a), each $f \in C_C(X)$ is $\mathcal{G}$-measurable; hence, $\mathcal{B}_0 \subset \mathcal{G}$. From (b), each compact G_δ is a Baire set, and so $\mathcal{G} \subset \mathcal{B}_0$. □

If D is discrete, then the Baire sets in βD are typically a proper sub-σ-algebra of the Borel sets in βD. If D is infinite, then a cardinality argument, similar to the one we used for $\beta \mathbb{N}$, would show that, for $t \in \beta D \backslash D$, the compact set $\{t\}$ is not a G_δ in βD.

Our next lemma explains why we never seemed to need the Baire sets before: On $\mathbb{R}$, or on a compact metric space, the Baire sets coincide with the Borel sets.

Lemma 16.6. *Let X be a second countable, locally compact Hausdorff space. Then,*

(i) *Every open set in X is a countable union of compact sets.*

(ii) *Every compact set in X is a G_δ.*

(iii) *The Baire and Borel σ-algebras on X coincide.*

Proof. Because X is locally compact, it has a base of compact neighborhoods. Since X is second countable, we can find a base consisting of only countably many such compact neighborhoods. Thus, (i) follows. And (ii) clearly follows from (i) by taking complements. Finally, (iii) follows from (i), (ii), and Lemma 16.5 (c). □

For good measure, here's another example:

Example. On an uncountable discrete space D, the Baire sets are a proper sub-σ-algebra of the Borel sets.

Proof. Since D is discrete, the only compact subsets of D are finite. It follows that the Baire σ-algebra on D is the σ-algebra generated by the singletons:

$$\mathcal{B}_0 = \{E : E \text{ or } E^c \text{ is countable}\}.$$

Because we can write D as the union of two disjoint subsets, each having the same cardinality as D, we can obviously find an (open) subset of D that is not a Baire set. $\quad\square$

The Dual of ℓ_∞

We are now well prepared to compute the dual of $\ell_\infty(D)$. As you might imagine, the continuous linear functionals on $\ell_\infty(D)$ should look like integration against some measure on βD. As a first step in this direction, we introduce the space ba (2^D), the collection of all *finitely additive* signed measures of finite variation on 2^D supplied with the norm $\|\mu\| = |\mu|(D)$, where $|\mu|$ is the total variation of μ. The name "ba" stands for "bounded (variation and finitely) additive."

Recall that the total variation of μ is defined by

$$|\mu|(E) = \sup\left\{\sum_{i=1}^n |\mu(E_i)| : E_1, \ldots, E_n \text{ disjoint}, E \subset \bigcup_{i=1}^n E_i\right\},$$

which applies equally well to finitely additive measures.

But what is meant by integration against such measures? Well, if $\mu \in$ ba (2^D), then $\int_D f \, d\mu$ is well defined and linear for simple (finitely-many-valued) functions f. Now, given a simple function f, write $f = \sum_{i=1}^n a_i \chi_{E_i}$, where $E_1, \ldots, E_n$ are disjoint and partition D. Then,

$$\left|\int_D f \, d\mu\right| = \sum_{i=1}^n |a_i \mu(E_i)| \le \|f\|_\infty \sum_{i=1}^n |\mu(E_i)| \le \|f\|_\infty \|\mu\|.$$

Thus, $\int_D f \, d\mu$ defines a bounded linear functional on the subspace of simple functions in $\ell_\infty(D)$. Since the simple functions are dense in $\ell_\infty(D)$, this means that $\int_D f \, d\mu$ extends unambiguously to all $f \in \ell_\infty(D)$. This unique linear extension to $\ell_\infty(D)$ is what we mean by $\int_D f \, d\mu$. With this understanding, our work is half done!

Theorem 16.7. $\ell_\infty(D)^* = $ ba (2^D) *isometrically.*

Proof. As we've just seen, each $\mu \in$ ba (2^D) defines a functional $x^* \in \ell_\infty(D)^*$ by setting $x^*(f) = \int_D f \, d\mu$, for f simple, and extending to all of $\ell_\infty(D)$. And, as the calculation in the previous paragraph shows, $\|x^*\| \le \|\mu\|$.

Next, the "hard" direction: Let $x^* \in \ell_\infty(D)^*$, and define $\mu(E) = x^*(\chi_E)$, for $E \subset D$. Clearly, μ is finitely additive, and we just need to check that μ is of bounded variation.

Given disjoint subsets $E_1, \ldots, E_n$ of D, we have

$$\sum_{i=1}^{n} |\mu(E_i)| = \sum_{i=1}^{n} |x^*(\chi_{E_i})|$$

$$= \sum_{i=1}^{n} \varepsilon_i x^*(\chi_{E_i}), \quad \text{for some } \varepsilon_i = \pm 1$$

$$= x^* \left(\sum_{i=1}^{n} \varepsilon_i \chi_{E_i} \right), \quad \text{by linearity}$$

$$\leq \|x^*\|, \quad \text{since} \quad \left\| \sum_{i=1}^{n} \varepsilon_i \chi_{E_i} \right\|_\infty = 1.$$

Thus, $\|\mu\| = |\mu|(D) \leq \|x^*\|$. Also, by linearity, we have that $x^*(f) = \int_D f \, d\mu$ for any simple function f. Since both functionals are continuous and agree on a dense subspace of $\ell_\infty(D)$, we necessarily have $x^*(f) = \int_D f \, d\mu$ for all $f \in \ell_\infty(D)$.

Combining the two halves of our proof, we arrive at the conclusion that the correspondence $x^* \leftrightarrow \mu$ is a linear isometry between the spaces $\ell_\infty(D)$ and $\mathrm{ba}(2^D)$. $\square$

Please note that our proof actually shows something more: *The positive linear functionals in $\ell_\infty(D)^*$ correspond to positive measures in* $\mathrm{ba}(2^D)$.

The Riesz Representation Theorem for $C(\beta D)$

Although knowing the dual of $\ell_\infty(D)$ should be reward enough, we could hope for more from our result. It falls just short of the full glory of the Riesz representation theorem for $C(\beta D)$: Optimistically, we'd like to represent the elements of $C(\beta D)^*$ as *regular, countably additive measures* on βD. As you can imagine, we might want to explore the possibility of applying Carathéodory's extension theorem to the elements of $\mathrm{ba}(2^D)$. But notice, please, that the natural σ-algebra associated to integration on $C(\beta D)$ is the Baire σ-algebra – in this case Σ. The approach we'll take only supplies a Baire measure. It's a fact, however, that every Baire measure on a compact Hausdorff space is regular; moreover, there is a standard technique for extending a Baire measure to a unique, regular Borel measure (see [48, Chapter 7], for example).

Now although the elements of ba (2^D) are not typically countably additive, they do satisfy a somewhat weaker property. Given a sequence of disjoint sets (A_i) in D, we have

$$\left| \sum_{i=1}^{\infty} \mu(A_i) \right| \leq \sum_{i=1}^{\infty} |\mu(A_i)| \leq |\mu| \left(\bigcup_{i=1}^{\infty} A_i \right) \leq \|\mu\| < \infty.$$

Hence, if μ is nonnegative, then

$$\sum_{i=1}^{\infty} \mu(A_i) \leq \mu \left(\bigcup_{i=1}^{\infty} A_i \right).$$

If μ is not countably additive, then what could the sum $\sum_{i=1}^{\infty} \mu(A_i)$ miss that $\mu(\bigcup_{i=1}^{\infty} A_i)$ picks up? The answer comes from βD.

The fact that disjoint sets in D have disjoint closures in βD allows us to define a "twin" of μ on βD: We identify each subset A of D with $\overline{A} = \text{cl}_{\beta D} A$ in βD and we define $\overline{\mu}(\overline{A}) = \mu(A)$. The set function $\overline{\mu}$ is a finitely additive measure on $\mathcal{A}$, the algebra of clopen subsets of βD. We can use $\overline{\mu}$ to explain how μ might fall short of being countably additive.

If (A_i) is a sequence of disjoint subsets of D, then, in general,

$$\bigcup_{i=1}^{\infty} \overline{A}_i \quad \underset{\neq}{\subseteq} \quad \overline{\bigcup_{i=1}^{\infty} \overline{A}_i}.$$

Why? Because the union on the left is *open*, while the union on the right is *compact*; equality can only occur if all but finitely many of the A_i's are *empty*! Thus, in general, $\sum \overline{\mu}(\overline{A}_i) \leq \overline{\mu}(\bigcup \overline{A}_i)$; that is, μ fails to account for the *closure* of $\bigcup \overline{A}_i$.

But this same observation shows us that $\overline{\mu}$ is actually *countably additive* on $\mathcal{A}$, the algebra of clopen sets. Indeed, if $\bigcup \overline{A}_i$ is *clopen*, then it's actually a *finite* union and so must equal $\bigcup \overline{A}_i$. In the terminology of [48], $\overline{\mu}$ is a "premeasure" on $\mathcal{A}$. Thus, we can invoke Carathéodory's theorem to extend $\overline{\mu}$ to a (regular) countably additive measure on Σ, the σ-algebra generated by $\mathcal{A}$. The extension $\overline{\mu}$ will still satisfy $\overline{\mu}(\overline{A}) = \mu(A)$ whenever $A \subset D$, of course. In particular, if $f = \sum_{i=1}^{n} a_i \chi_{A_i}$ is a simple function based on disjoint subsets of D, then $\tilde{f} = \sum_{i=1}^{n} a_i \chi_{\overline{A}_i}$ is a simple function based on disjoint sets in $\mathcal{A}$, and we have

$$\int_D f \, d\mu = \sum_{i=1}^{n} a_i \mu(A_i) = \sum_{i=1}^{n} a_i \overline{\mu}(\overline{A}_i) = \int_{\beta D} \tilde{f} \, d\overline{\mu}.$$

What this means is that we can represent the elements of $C(\beta D)^*$ as integration against regular, countably additive measures on Σ. If $T : \ell_{\infty}(D) \rightarrow$

$C(\beta D)$ is the canonical isometry, notice that T maps the characteristic function of a set A in D to the characteristic function of $\mathrm{cl}_{\beta D} A$ in βD. Thus, T maps simple functions based on sets in 2^D to simple functions based on clopen sets in $\mathcal{A}$. Now a functional $x^* \in C(\beta D)^*$ induces a functional $x^* \circ T$ on $\ell_\infty(D)$. Hence, there is a measure $\mu \in \mathrm{ba}\,(2^D)$ such that $x^*(Tf) = \int_D f\, d\mu$ for every $f \in \ell_\infty(D)$. If f is a simple function, then $Tf = \tilde{f}$ in the notation we used above, and so

$$x^*\big(\tilde{f}\big) = x^*(Tf) = \int_D f\, d\mu = \int_{\beta D} \tilde{f}\, d\bar{\mu}.$$

Since the simple functions based on clopen sets are uniformly dense in $C(\beta D)$, we must have $x^*(g) = \int_{\beta D} g\, d\bar{\mu}$ for every $g \in C(\beta D)$. That is, we've arrived at the Riesz representation theorem for $C(\beta D)$. In symbols,

$$C(\beta D)^* = \ell_\infty(D)^* = \mathrm{ba}\,(2^D) = \mathrm{rca}\,(\Sigma),$$

where $\mathrm{rca}\,(\Sigma)$ denotes the space of regular, countably additive measures on Σ.

Theorem 16.8. *Given a continuous linear functional $x^* \in C(\beta D)^*$, there exists a unique signed measure μ on the Baire sets in βD such that*

$$x^*(f) = \int_{\beta D} f\, d\mu \qquad \text{for all } f \in C(\beta D).$$

Moreover, $\|x^*\| = |\mu|(\beta D)$.

Although we have not addressed uniqueness in this representation, it follows the usual lines. That we have equality of norms for the representing measure can again be attributed to the fact that the simple functions are dense in $C(\beta D)$.

Now we're ready to apply this result to the problem of representing the elements of $C(K)^*$ as integration against Baire measures on K. To begin, let K be a compact Hausdorff space. Next, we choose a discrete space D and a continuous, onto map $\varphi : \beta D \to K$. Then, as you'll recall, the map $f \mapsto f \circ \varphi$ defines a linear isometry from $C(K)$ into $C(\beta D)$; in other words, each continuous $f : K \to \mathbb{R}$ "lifts" to βD by way of $f \circ \varphi : \beta D \to \mathbb{R}$. Thus, each $x^* \in C(K)^*$ extends to a functional $y^* \in C(\beta D)^*$ satisfying $x^*(f) = y^*(f \circ \varphi)$, for all $f \in C(K)$, and $\|y^*\| = \|x^*\|$. That is, y^* is a Hahn–Banach extension of the functional $g \mapsto x^*(g \circ \varphi^{-1})$ defined on the image of $C(K)$ in $C(\beta D)$. (And it's not hard to check that a positive functional has a positive extension.)

From the Riesz representation theorem for $C(\beta D)$, we can find a measure μ defined on the Baire sets in βD such that

$$y^*(g) = \int_{\beta D} g \, d\mu \qquad \text{for all } g \in C(\beta D).$$

Hence

$$x^*(f) = y^*(f \circ \varphi) = \int_{\beta D} (f \circ \varphi) \, d\mu$$

$$= \int_K f \, d\,(\mu \circ \varphi^{-1}) \qquad \text{for all } f \in C(K),$$

and $\nu(A) = \mu(\varphi^{-1}(A))$ defines a Baire measure on K.

The uniqueness of μ follows from the regularity of μ and Urysohn's lemma; it requires checking that $\int_K f \, d\mu = 0$ for all $f \in C(K)$ forces $\mu \equiv 0$. The details are left as an exercise.

Notes and Remarks

Our presentation in this chapter borrows heavily from Garling's paper [53], but see also Diestel [33], Gillman and Jerison [55], Hartig [65], Holmes [70], Kelley [84], and Yosida and Hewitt [149]. An approach to Riesz's theorem that would have pleased Riesz can be found in Dudley's book [39].

Exercises

1. Show that X is extremally disconnected if and only if disjoint open sets in X have disjoint closures.

2. Let X be completely regular. Show that X is extremally disconnected if and only if βX is extremally disconnected.

3. If D is an infinite discrete space, prove that βD is not sequentially compact and, hence, not metrizable.

4. Let K be a compact Hausdorff space and let $\mathcal{B}$ and $\mathcal{B}_0$ denote the Borel and Baire σ-algebras on K, respectively. Prove that $\mathcal{B}_0 \subset \mathcal{B}$.

Appendix
Topology Review

We denote a topological space by $(X, \mathcal{T})$, where X is a set and $\mathcal{T}$ is a *topology* on X. That is, $\mathcal{T}$ is a collection of subsets of X, called *open sets*, satisfying (i) $\emptyset, X \in \mathcal{T}$, (ii) $U, V \in \mathcal{T} \implies U \cap V \in \mathcal{T}$, and (iii) $\mathcal{A} \subset \mathcal{T} \implies \bigcup \mathcal{A} \in \mathcal{T}$. The *closed sets* in X are the complements of the open sets; that is, a subset E of X is closed if E^c is open. As shorthand, reference to the topology $\mathcal{T}$ is often only implicit as in the phrase: "Let X be a topological space."

Every set X supports at least two topologies. Indeed, it's easy to check that $\{\emptyset, X\}$ is a topology on X, called the *indiscrete topology*, and that $\mathcal{P}(X)$, the power set of X, is a topology on X, called the *discrete topology*. We say that X is a *discrete space* if X is endowed with its discrete topology. Please note that every subset of a discrete space is both open and closed.

Once we have the notion of an open set, we can consider continuous functions between topological spaces: A function $f : X \to Y$ from a topological space X to a topological space Y is *continuous* if $f^{-1}(U)$ is open in X whenever U is open in Y. The collection of all continuous functions from X into Y is denoted by $C(X; Y)$. In case $Y = \mathbb{R}$, we shorten $C(X; \mathbb{R})$ to $C(X)$. Various subsets of $C(X)$, such as $C_b(X)$, $C_C(X)$, and so on, have the same meaning as in the introductory chapter.

We can also consider compact sets: A subset K of a topological space X is said to be *compact* if every covering of K by open sets admits a finite subcover; that is, K is compact if, given any collection of open sets $\mathcal{U}$ satisfying $\bigcup\{V : V \in \mathcal{U}\} \supset K$, we can always reduce to finitely many sets $V_1, \ldots, V_n \in \mathcal{U}$ with $V_1 \cup \cdots \cup V_n \supset K$. It's easy to see that compact sets are necessarily closed.

Separation

Recall that a topological space $(X, \mathcal{T})$ is said to be *Hausdorff* if distinct points in X can always be separated by disjoint open sets; that is, given $x \neq y \in X$, we can find disjoint sets $U, V \in \mathcal{T}$ such that $x \in U$ and $y \in V$. Please note that, in a Hausdorff space, each singleton $\{x\}$ is a closed set. More generally, each compact subset of a Hausdorff space is closed. Just as with metric spaces, "closed" will mean "closed under limits" (we'll make this precise shortly), whereas "compact" will mean "limits exist in abundance." We would prefer those existential limits to land back in our compact set, and so it's helpful to know that a compact set is closed. For this reason (among others), it's easier to do analysis in a Hausdorff space. In fact, it's quite rare for an analyst to encounter (or even consider!) a topological space that fails to be Hausdorff. Henceforth, we will assume that *ALL* topological spaces are Hausdorff. Be forewarned, though, that this blanket assumption may also mean that a few of our definitions are fated to be nonstandard.

Metric spaces and compact Hausdorff spaces enjoy an even stronger separation property; in either case, disjoint closed sets can always be separated by disjoint open sets. A Hausdorff topological space is said to be *normal* if it has this property: Given disjoint closed sets E, F in X, there are disjoint open sets U, V in $\mathcal{T}$ such that $E \subset U$ and $F \subset V$.

Normality has two other characterizations, each important in its own right. The first is given by Urysohn's lemma: In a normal topological space, disjoint closed sets can be *completely separated*. That is, if E and F are disjoint closed sets in a normal space X, then there is a continuous function $f \in C(X; [0, 1])$ such that $f = 0$ on E while $f = 1$ on F. The second is Tietze's extension theorem: If E is a closed subset of a normal space X, then each continuous function $f \in C(E; [0, 1])$ extends to a continuous function $\tilde{f} \in C(X; [0, 1])$ on all of X. Urysohn's lemma and Tietze's theorem are each equivalent to normality. The interval $[0, 1]$ can be replaced in either statement by an arbitrary interval $[a, b]$. Further, it follows from Tietze's theorem that if E is a closed subset of a normal space X, then every $f \in C(E)$ extends to an element of $C(X)$ (simply by composing f with a suitable homeomorphism from $\mathbb{R}$ into $(0, 1)$).

Locally Compact Hausdorff Spaces

If X has "enough" compact neighborhoods, then $C_C(X)$ will have enough functions to take the place of $C(X)$ in certain situations. In this context, "enough" means that X should be *locally compact*. A locally compact Hausdorff space is one in which each point has a compact neighborhood; that is, given $x \in X$, there is an open set U containing x such that $\overline{U}$ is compact. It's an easy exercise to show that a locally compact Hausdorff space has a wealth of compact neighborhoods in the following sense: Given $K \subset U \subset X$, where K is compact and U is open, there is an open set V with compact closure such that $K \subset V \subset \overline{V} \subset U$. This observation (along with a bit of hard work!) leads to locally compact versions of both Urysohn's lemma and Tietze's extension theorem. Here's how they now read (please note that $C_C(X)$ is used in place of $C(X)$ in each case): Let X be a locally compact Hausdorff space, and let K be a compact subset of X. (Urysohn) If F is a closed set disjoint from K, then there is an $f \in C_C(X; [0, 1])$ such that $f = 0$ on K while $f = 1$ on F. (Tietze) Each element of $C(K)$ extends to an element of $C_C(X)$.

An alternate approach is to consider the *one-point compactification* of X. The one-point compactification X^* of a topological space X is defined to be the space $X^* = X \cup \{\infty\}$, where ∞ is a distinguished point appended to X and where we define a topology on X^* by taking the neighborhoods of ∞ to be sets of the form $\{\infty\} \cup U$, where U^c is compact in X. It's easy to see that X^* is a compact space that contains X as an open, dense subset (this is what it means to be a *compactification* of X). It is likewise easy to see that X^* is Hausdorff precisely when X is locally compact and Hausdorff. Thus, if X is a locally compact Hausdorff space, then X is a dense, open subset of the compact Hausdorff (hence normal) space X^*. Consequently, we can now take advantage of such niceties as Urysohn's lemma and Tietze's extension theorem in X^* and then simply translate these properties to X. In particular, the locally compact versions of Urysohn's lemma and Tietze's theorem, stated above, are direct consequences of considering the "full" versions in X^* and then "cutting back" to X.

If X is locally compact, then the completion of $C_C(X)$ under the sup norm is the space $C_0(X)$, the functions in $C(X)$ that "vanish at infinity"; that is, those $f \in C(X)$ for which the set $\{|f| \geq \varepsilon\}$ is compact for every $\varepsilon > 0$. Clearly, $C_0(X)$ is a Banach space (and a Banach algebra) under the sup norm. The phrase "vanish at infinity" becomes especially meaningful if we consider X^*, the one-point compactification of X. In this setting, the space $C_0(X)$ is (isometrically) the collection of functions in $C(X^*)$ that are zero at ∞.

It may come as a surprise to learn that the *discrete spaces* are a very important class of locally compact spaces. In the case of a discrete space D, the various spaces of continuous functions are often given different names. For example, since every function $f : D \to \mathbb{R}$ is continuous, the space $C_b(D)$ is simply the collection of *all* bounded functions on D, and this is often written as $\ell_\infty(D)$ in analogy with the sequence space $\ell_\infty = \ell_\infty(\mathbb{N})$. The space $C_C(D)$ is the collection of functions with finite support in $\ell_\infty(D)$, and the space $C_0(D)$ is often written as $c_0(D)$, again in keeping with the sequence space notation $c_0 = C_0(\mathbb{N})$. As a last curiosity, notice that the space c of all convergent sequences is just a renaming of the space $C(\mathbb{N} \cup \{\infty\})$.

Weak Topologies

A familiar game is to describe the continuous functions on X *after* we've been handed a topology on X. But the inverse procedure is just as common and perhaps even more useful. In other words, given a collection of functions $\mathcal{F}$ from a *set* X to some fixed topological space Y, can we construct a topology on X under which each element of $\mathcal{F}$ will be continuous? If X is given the discrete topology, then every function from X to Y is continuous, whereas if X is given the trivial, or indiscrete topology, then only constant functions are continuous. We typically want something in between. In fact, we'd like to know if there is a *smallest* (or *weakest*) topology that makes each element of $\mathcal{F}$ continuous. As we'll see, the answer is yes and follows easily from an important bit of machinery that provides for the construction of topologies having certain predetermined open sets.

Lemma A.1 (Subbasis Lemma). *Suppose that X is a set and that $\mathcal{S}$ is a collection of subsets of X. Then, there is a smallest topology $\mathcal{T}$ on X containing $\mathcal{S}$. Moreover, $\mathcal{S}' = \{\emptyset, X\} \cup \mathcal{S}$ forms a subbase for $\mathcal{T}$. In other words, the sets of the form $S_1 \cap \cdots \cap S_n$, where $S_i \in \mathcal{S}'$ for $i = 1, \ldots, n$, are a base for $\mathcal{T}$.*

Proof. Let $\mathcal{T}_1$ denote the intersection of all topologies on X containing $\mathcal{S}$ (or $\mathcal{S}'$). It's easy to see that $\mathcal{T}_1$ is itself a topology on X containing $\mathcal{S}$ (as well as $\mathcal{S}'$), and clearly, $\mathcal{T}_1$ is the smallest such topology. Consequently, $\mathcal{T}_1$ also contains the collection $\mathcal{T}_2$, which we define to be the set of all possible unions of sets of the form $S_1 \cap \cdots \cap S_n$, where $n \geq 1$ and where $S_i \in \mathcal{S}'$ for $i = 1, \ldots, n$. All that remains is to show that $\mathcal{T}_1 = \mathcal{T}_2$. But since $\mathcal{T}_2$ contains $\mathcal{S}$, it suffices to show that $\mathcal{T}_2$ is a topology on X. In fact, the only detail that we need to check is that $\mathcal{T}_2$ is closed under finite intersections (since it's obviously closed under arbitrary unions). Here goes: Let $U, V \in \mathcal{T}_2$ and let $x \in U \cap V$. Take $A_1, \ldots, A_n$ and $B_1, \ldots, B_m$ in $\mathcal{S}'$ such that $x \in A_1 \cap \cdots \cap A_n \subset U$ and $x \in B_1 \cap \cdots \cap B_m \subset V$. Then $x \in A_1 \cap \cdots \cap A_n \cap B_1 \cap \cdots \cap B_m \subset U \cap V$. That is, $U \cap V \in \mathcal{T}_2$. □

And how does Lemma A.1 help? Well, given a set of functions $\mathcal{F}$ from X into a topological space Y, take the smallest topology on X containing the sets

$$\mathcal{S} = \{f^{-1}(U) : f \in \mathcal{F}, \ U \text{ is open in } Y\}.$$

Because each element of $\mathcal{S}$ will be open in the new topology, each element of $\mathcal{F}$ will be continuous. This topology is usually referred to as the *the weak topology induced by $\mathcal{F}$*.

We don't really need the inverse image of every open set in Y; we could easily get by with just the inverse images of a collection of basic open sets, or even subbasic open sets. In particular, if $Y = \mathbb{R}$, then the collection of sets

$$N(x; f_1, \ldots, f_n; \varepsilon) = \{y \in X : |f_i(x) - f_i(y)| < \varepsilon, \ i = 1, \ldots, n\},$$

where $x \in X$, $f_1, \ldots, f_n \in \mathcal{F}$, and $\varepsilon > 0$, is a neighborhood base for the weak topology generated by $\mathcal{F}$.

If X carries the weak topology induced by a collection of functions $\mathcal{F}$ from X into Y, then it's easy to describe the continuous functions *into* X; note that $f : Z \rightarrow X$ is continuous if and only if $g \circ f : Z \rightarrow Y$ is continuous for every $g \in \mathcal{F}$.

Finally, it's worth pointing out that our construction of weak topologies in no way requires a fixed range space Y. In particular, given a collection of functions $(f_\alpha)_{\alpha \in A}$, where f_α maps X into a topological space Y_α, we can easily apply the subbasis lemma to find the smallest topology on X under which each f_α is continuous. In this setting, we consider the topology generated by the collection

$$\mathcal{S} = \{f_\alpha^{-1}(U_\alpha) : U_\alpha \text{ is open in } Y_\alpha \text{ for } \alpha \in A\}.$$

Product Spaces

The subbasis lemma readily adapts to more elaborate applications. The product (or Tychonoff) topology provides an excellent example of such an adaptation. First, recall that the (Cartesian) product of a collection of (nonempty) sets $(Y_\alpha)_{\alpha \in A}$ is defined to be the set of all functions $f : A \rightarrow \cup_{\alpha \in A} Y_\alpha$ satisfying $f(\alpha) \in Y_\alpha$; the product space is written $\prod_{\alpha \in A} Y_\alpha$. If we identify an element f of the product space with its range $(f(\alpha))_{\alpha \in A}$, then we recover the familiar notion that the product space consists of "tuples," where the βth "coordinate" of each tuple is to be an element of Y_β. We also have the familiar coordinate projections $\pi_\beta : \prod_{\alpha \in A} Y_\alpha \rightarrow Y_\beta$ defined by the formula $\pi_\beta(f) = f(\beta)$ (or, $\pi_\beta((f(\alpha))_{\alpha \in A}) = f(\beta)$). If each Y_α is the same set Y, we usually write the product space as Y^A, the set of all functions from A into Y.

In case each Y_α is a topological space, we topologize the product space $\prod_{\alpha \in A} Y_\alpha$ by giving it the weak topology induced by the π_α's. That is, we define *the product topology* to be the smallest topology under which all of the coordinate projections are continuous. In terms of the subbasis lemma, this means that each of the sets $\pi_\alpha^{-1}(U_\alpha)$, where U_α is open in Y_α and $\alpha \in A$, is a subbasic open set in the product. The basic open sets in the product are, of course, finite intersections of these.

This particular choice for a topology on the product space results in several useful consequences. For example, a function $\varphi : X \to \prod_{\alpha \in A} Y_\alpha$, from a topological space X into a product space, is continuous if and only if each of its "coordinates" $\pi_\alpha \circ \varphi : X \to Y_\alpha$ is continuous. We'll see other benefits of the product topology next.

Nets

Now this being analysis (or had you forgotten?), we need a valid notion of limit, or convergence, in a general topological space. An easy choice, from our point of view, is to consider nets. The reader who is unfamiliar with nets would be well served by thinking of a net as a "generalized sequence." We start with a *directed set* D; that is, D is equipped with a binary relation $\leq$ satisfying (i) $\lambda \leq \lambda$ for all $\lambda \in D$; (ii) if $\lambda \leq \mu$ and $\mu \leq \nu$, then $\lambda \leq \nu$; and (iii) given any $\lambda, \mu \in D$, there is some $\nu \in D$ with $\lambda \leq \nu$ and $\mu \leq \nu$. Several standard examples come to mind. $\mathbb{N}$, with its usual order, is a directed set. The set of all finite subsets of a fixed set is directed by inclusion (i.e., $A \leq B$ if $A \subset B$). The set of all neighborhoods of a fixed point in any topological space is directed by *reverse* inclusion (i.e., $A \leq B$ if $A \supset B$). And so on. As usual, we also write $\lambda \geq \mu$ to mean $\mu \leq \lambda$.

Now, a *net* in a set X is any function into X whose *domain* is a directed set. A sequence, recall, is a function with domain $\mathbb{N}$ and so is also a net. Just as with sequences, though, we typically identify a net with its range. In other words, we would denote a net in X by simply writing $(x_\lambda)_{\lambda \in D}$, where D is a directed set and where each $x_\lambda \in X$.

In defining convergence for nets, we just tailor the terminology we already use for sequences. For example, we say that a net $(x_\lambda)_{\lambda \in D}$ is *eventually* in the set A if, for some $\mu \in D$, we have $\{x_\lambda : \lambda \geq \mu\} \subset A$. And $(x_\lambda)_{\lambda \in D}$ is *frequently* in A if, given any $\lambda \in D$, there is some $\mu \in D$ with $\mu \geq \lambda$ such that $x_\mu \in A$. Finally, a net $(x_\lambda)_{\lambda \in D}$ in a topological space X *converges* to the point $x \in X$ if $(x_\lambda)_{\lambda \in D}$ is eventually in each neighborhood of x. As with sequences, we use the shorthand $x_\lambda \to x$ in this case.

Many topological properties can be characterized in terms of convergent nets. Indeed, sequential characterizations used in metric spaces can typically be directly translated into this new language of nets. Here's an easy example: A set E in a topological space X is closed if and only if each net (x_λ) in E that converges in X actually converges to a point of E. On the one hand, suppose that E is closed and let (x_λ) be a net in E converging to $x \in X$. If $x \in E^c$, an open set, then we would have to have (x_λ) eventually in E^c, which is an impossiblility. On the other hand, suppose that each convergent net from E converges to a point in E. Now let $x \in E^c$ and suppose that for every neighborhood U of x there is some point $x_U \in U \cap E$. If we direct the neighborhoods of x by reverse inclusion, then we've just constructed a net (x_U) in E that converges to a point x not in E, yielding a contradiction. Thus, some neighborhood U of x is completely contained in E^c; that is, E^c is open.

Another example that's easy to check: A function $f : X \to Y$ between topological spaces is continuous if and only if the net $(f(x_\lambda))$ converges to $f(x) \in Y$ whenever the net (x_λ) converges to $x \in X$. Suppose first that f is continuous. Let $x_\lambda \to x$ in X and let U be a neighborhood of $f(x)$ in Y. Then $f^{-1}(U)$ is a neighborhood of x in X and, hence, (x_λ) is eventually in $f^{-1}(U)$. Consequently, $(f(x_\lambda))$ is eventually in U. That is, $(f(x_\lambda))$ converges to $f(x)$. Next suppose that $f(x_\lambda) \to f(x)$ whenever $x_\lambda \to x$. Let E be a closed set in Y and let (x_λ) be a net in $f^{-1}(E)$ that converges to $x \in X$. Then $(f(x_\lambda))$ is a net in E that converges to $f(x) \in Y$. Hence, $f(x) \in E$ or $x \in f^{-1}(E)$. Thus, $f^{-1}(E)$ is closed and so f is continuous.

Nets will prove especially useful in arguments involving weak topologies. If X carries the weak topology induced by a family of functions $\mathcal{F}$, it follows that a net (x_λ) converges to $x \in X$ if and only if $(f(x_\lambda))$ converges to $f(x)$ for each $f \in \mathcal{F}$. (Why?)

Now since the product topology is nothing more than a weak topology, our latest observation takes on a very simple guise in a product space. A net (f_λ) in a product space $\prod_{\alpha \in A} Y_\alpha$ converges to f if and only if (f_λ) converges "coordinatewise"; that is, if and only if $(f_\lambda(\alpha))$ converges to $f(\alpha)$ for each $\alpha \in A$. For this reason, the product topology is sometimes called *the topology of pointwise convergence*. Beginning to sound like analysis?

The only potential hardship with nets is that the notion of a *subnet* is a bit more complicated than the notion of a subsequence. But we're in luck: We will have no need for subnets, and so we may blissfully ignore their intricacies.

Notes and Remarks

There are many excellent books on topology (or on topology for analysts) that will provide more detail than we have given here. See, for example, Folland [48, Chapter 4], Jameson [76], Kelley [84], Kelley and Namioka [85], Köthe [86], Simmons [138], or Willard [146].

References

[1] R. A. Adams, *Sobolev Spaces*, Pure and Applied Mathematics v. 65, Academic Press, New York, 1975.

[2] L. Alaoglu, Weak topologies of normed linear spaces, *Annals of Mathematics* **41**(1940), 252–267.

[3] C. Aliprantis and O. Burkinshaw, *Positive Operators*, Academic Press, Orlando, 1985.

[4] T. Ando, Contractive projections in L_p spaces, *Pacific Journal of Mathematics* **17**(1966), 391–405.

[5] S. Banach, "Sur les opérations dans les ensembles abstraits et leur application aux équations intégrals," *Fundamenta Mathematicae* **3**(1922), 133–181.

[6] S. Banach, *Théorie des Opérations Linéaires*, 2nd ed., Chelsea, New York, 1955.

[7] S. Banach, *Theory of Linear Operations*, North-Holland Mathematical Library v. 38, North-Holland, New York 1987.

[8] S. Banach and S. Mazur, Zur Theorie der linearen Dimension, *Studia Mathematica* **4**(1933), 100–112.

[9] S. Banach and S. Saks, Sur la convergence forte dans le champ L_p, *Studia Mathematica* **2**(1930), 51–57.

[10] S. Banach and H. Steinhaus, Sur le principle de la condensation de singularités, *Fundamenta Mathematicae* **9**(1927), 50–61.

[11] B. Beauzamy, *Introduction to Banach Spaces and their Geometry*, North-Holland Mathematics Studies v. 68, North-Holland, New York 1982.

[12] C. Bennett and R. Sharpley, *Interpolation of Operators*, Pure and Applied Mathematics v. 29, Academic Press, Orlando, 1988.

[13] M. Bernkopf, The development of function spaces with particular reference to their origins in integral equation theory, *Archive for History of Exact Sciences* **3**(1966), 1–96.

[14] M. Bernkopf, A history of infinite matrices, *Archive for History of Exact Sciences* **4**(1967–68), 308–358.

[15] C. Bessaga and A. Pełczyński, On bases and unconditional convergence of series in Banach spaces, *Studia Mathematica* **17**(1958), 151–164.

[16] C. Bessaga and A. Pełczyński, Properties of bases in B_0 spaces [in Polish], *Prace Matematyczne* **3**(1959), 123–142.

[17] R. P. Boas, Some uniformly convex spaces, *Bulletin of the American Mathematical Society* **46**(1940), 304–311.

[18] B. Bollobás, *Linear Analysis*, Cambridge University Press, New York, 1990.

[19] N. L. Carothers, *Real Analysis*, Cambridge University Press, New York, 2000.
[20] P. G. Casazza, The Schroeder–Bernstein property for Banach spaces, in *Analysis at Urbana*, E. Berkson, T. Peck, J. J. Uhl., eds., 2 vols., London Mathematical Society Lecture Notes **137–138**, Cambridge University Press, New York 1986–87.
[21] P. G. Casazza, The Schroeder–Bernstein property for Banach spaces, in *Banach Space Theory*, Bor-Luh Lin, ed., Contemporary Mathematics **85**, American Mathematical Society, 1989.
[22] P. G. Casazza, Some questions arising from the homogeneous Banach space problem, in *Banach Spaces*, Bor-Luh Lin and W. B. Johnson, eds., Contemporary Mathematics **144**, American Mathematical Society, 1993.
[23] P. G. Casazza, "Classical Sequences in Banach Spaces," by Sylvie Guerre-Delabrière, *Bulletin of the American Mathematical Society* (N.S.) **30**(1994), 117–124; book review.
[24] E. Čech, On bicompact spaces, *Annals of Mathematics* **38**(1937), 823–844.
[25] J. A. Clarkson, Uniformly convex spaces, *Transactions of the American Mathematical Society* **40**(1936), 396–414.
[26] J. B. Conway, *A Course in Functional Analysis*, 2nd ed., Graduate Texts in Mathematics 96, Springer-Verlag, New York, 1990.
[27] T. A. Cook, On normalized Schauder bases, *The American Mathematical Monthly* **77**(1970), 167.
[28] M. M. Day, Strict convexity and smoothness of normed spaces, *Transactions of the American Mathematical Society* **78**(1955), 516–528.
[29] M. M. Day, *Normed Linear Spaces*, Ergebnisse der Mathematik und ihrer Grenzgebiete bd. 21, 3rd ed., Springer-Verlag, New York, 1973.
[30] C. DeVito, *Functional Analysis*, Academic Press, New York 1978.
[31] J. Diestel, *Geometry of Banach Spaces – Selected Topics*, Lecture Notes in Mathematics **485**, Springer-Verlag, New York, 1975.
[32] J. Diestel, Review of *Théories des Opérations Linéaires* by Stephan Banach, *Mathematical Intelligencer* **4**(1982), 45–48.
[33] J. Diestel, *Sequences and Series in Banach Spaces*, Graduate Texts in Mathematics 92, Springer-Verlag, New York, 1984.
[34] J. Diestel and J. J. Uhl, *Vector Measures* American Mathematical Society, Providence, 1977.
[35] S. J. Dilworth, *Lecture Notes on Banach Space Theory*, unpublished.
[36] J. Dieudonné, *History of Functional Analysis*, North-Holland, New York 1981.
[37] J. Dixmier, Sur un théorème de Banach, *Duke Mathematical Journal* **15**(1948), 1057–1071.
[38] R. G. Douglas, Contractive projections on an L_1 space, *Pacific Journal of Mathematics* **15**(1965), 443–462.
[39] R. M. Dudley, *Real Analysis and Probability*, Wadsworth & Brooks/Cole, Pacific Grove, California, 1989.
[40] N. Dunford and A. Morse. Remarks on the preceding paper of James A. Clarkson, *Transactions of the American Mathematical Society* **40**(1936), 415–420.
[41] N. Dunford and B. J. Pettis, Linear operations on summable functions, *Transactions of the American Mathematical Society* **47**(1940), 323–392.
[42] N. Dunford and J. Schwartz, *Linear Operators*, Part I, Wiley Interscience, New York, 1958.

[43] P. Duren, Theory of H^p Spaces, Pure and Applied Mathematics 38, Academic Press, New York, 1970.

[44] A. Dvoretzky, Some results on convex bodies and Banach spaces, in *Proceedings of the International Symposium on Linear Spaces*, Jerusalem (1961), 123–160.

[45] A. Dvoretzky and C. A. Rogers, Absolute and unconditional convergence in normed linear spaces, *Proceedings of the National Academy of Sciences USA* **36**(1950), 192–197.

[46] P. Enflo, A counterexample to the approximation property in Banach spaces, *Acta Mathematica* **130**(1973), 309–317.

[47] K. Fan and I. Glicksberg, Some geometric properties of the spheres in a normed linear space, *Duke Journal of Mathematics* **25**(1958), 553–568.

[48] G. Folland, *Real Analysis*, Wiley, New York, 1984.

[49] M. Fréchet, *Les Espaces Abstraits*, Gauthier-Villars, Paris, 1928.

[50] K. O. Friedrichs, On Clarkson's inequalities, *Communications in Pure and Applied Mathematics* **23**(1970), 603–607.

[51] I. S. Gál, On sequences of operations in complete vector spaces, *The American Mathematical Monthly* **60**(1953), 527–538.

[52] V. Gantmacher, Über schwache totalstetige Operatoren, *Matematicheskii Sbornik* (N.S.) **7**(1940), 301–308.

[53] D. J. H. Garling, A "short" proof of the Riesz representation theorem, *Proceedings of the Cambridge Philosophical Society* **73**(1973), 459–460.

[54] D. J. H. Garling, *Lecture Notes on Probability in Banach Spaces*, unpublished.

[55] L. Gillman and M. Jerison, *Rings of Continuous Functions*, Graduate Texts in Mathematics 43, Springer-Verlag, New York, 1976.

[56] C. Goffman and G. Pedrick, *First Course in Functional Analysis*, 2nd ed., Chelsea, New York, 1983.

[57] S. Goldberg, A simple proof of a theorem concerning reflexivity, *The American Mathematical Monthly* **67**(1960), 1004.

[58] W. T. Gowers, A solution to Banach's Hyperplane Problem, *Bulletin of the London Mathematical Society* **26**(1994), 523–530.

[59] W. T. Gowers, A solution to the Schroeder–Bernstein problem for Banach spaces, *Bulletin of the London Mathematical Society* **28**(1996), 297–304.

[60] W. T. Gowers and B. Maurey, The unconditional basic sequence problem, *Journal of the American Mathematical Society* **6**(1993), 851–874.

[61] W. T. Gowers and B. Maurey, Banach spaces with small spaces of operators, *Mathematische Annalen* **307**(1997), 543–568.

[62] A. Grothendieck, Sur les applications linéaires faiblement compactes d'espaces du type $C(K)$, *Canadian Journal of Mathematics* **5**(1953), 129–173.

[63] U. Haagerup, Les meilleures constantes de l'inégalité de Khintchine, *Comptes Rendus hebdomadaires des Séances de l'Académie des Sciences, Paris* **286**(1978), 259–262.

[64] G. H. Hardy, J. E. Littlewood, and G. Pólya, *Inequalities*, 2nd ed., Cambridge University Press, New York, 1983.

[65] D. Hartig, The Riesz representation theorem revisited, *The American Mathematical Monthly* **90**(1983), 277–280.

[66] F. Hausdorff, *Grundzüge der Mengenlehre*, Von Veit, Leipzig, 1914; 3rd ed., Von Veit, Leipzig, 1937, published in English as *Set Theory*, 3rd English ed., Chelsea, New York, 1978.

[67] F. Hausdorff, Zur Theorie der linearen metrischen Räume, *Journal für die Reine und Angewandte Mathematik* **167**(1932), 294–311.

[68] J. Hennefeld, A nontopological proof of the uniform boundedness theorem, *The American Mathematical Monthly* **87**(1980), 217.

[69] E. Hewitt and K. Stromberg, *Real and Abstract Analysis*, Graduate Texts in Mathematics 25, Springer-Verlag, New York, 1965.

[70] R. B. Holmes, *Geometric Functional Analysis*, Graduate Texts in Mathematics 24, Springer-Verlag, New York, 1975.

[71] R. A. Hunt and G. Weiss, The Marcinkiewicz interpolation theorem, *Proceedings of the American Mathematical Society* **15**(1964), 996–998.

[72] R. C. James, Bases and reflexivity of Banach spaces, *Annals of Mathematics* **52**(1950), 518–527.

[73] R. C. James, A non-reflexive Banach space isometric with its second conjugate space, *Proceedings of the National Academy of Sciences USA* **37**(1951), 174–177.

[74] R. C. James, Uniformly non-square Banach spaces, *Annals of Mathematics* **80**(1964), 542–550.

[75] R. C. James, Bases in Banach spaces, *The American Mathematical Monthly* **89**(1982), 625–640.

[76] G. J. O. Jameson, *Topology and Normed Spaces*, A Halsted Press Book, Wiley, New York, 1974.

[77] G. J. O. Jameson, Whitley's technique and K_δ-subspaces of Banach spaces, *The American Mathematical Monthly* **84**(1977), 459–461.

[78] R. D. Järvinen, *Finite and Infinite Dimensional Linear Spaces* Marcel Dekker, New York 1981.

[79] J. L. W. V. Jensen, Sur les fonctions convexes et les inégalités entre les valeurs moyennes, *Acta Mathematica* **30**(1906), 175–193.

[80] W. B. Johnson, Complementably universal separable Banach spaces: An application of counterexamples to the approximation problem, in *Studies in Functional Analysis*, R. G. Bartle, ed., Mathematical Association of America, Washington, D.C. 1980.

[81] W. B. Johnson, B. Maurey, G. Schechtman, and L. Tzafriri, *Symmetric Structures in Banach Spaces*, *Memoirs of the American Mathematical Society* **217**, American Mathematical Society, 1979.

[82] M. I. Kadec and A. Pełczyński, Bases, lacunary sequences and complemented subspaces of L_p, *Studia Mathematica* **21**(1962), 161–176.

[83] M. Kanter, Stable laws and the imbedding of L_p spaces, *The American Mathematical Monthly* **80**(1973), 403–407.

[84] J. L. Kelley, *General Topology*, Graduate Texts in Mathematics 27, Springer-Verlag, New York, 1955.

[85] J. L. Kelley and I. Namioka, *Topological Vector Spaces*, Graduate Texts in Mathematics 36, Springer-Verlag, New York, 1976.

[86] G. Köthe, *Topological Vector Spaces I*, Die Grundlehren der mathematischen Wissenschaften bd. 159, Springer-Verlag, New York, 1969.

[87] H. E. Lacey, The Hamel dimension of any infinite dimensional separable Banach space is c, *The American Mathematical Monthly* **80**(1973), 298.

[88] H. E. Lacey, *The Isometric Theory of Classical Banach Spaces*, Die Grundlehren der mathematischen Wissenschaften bd. 208, Springer-Verlag, New York 1974.

[89] J. Lamperti, The isometries of certain function spaces, *Pacific Journal of Mathematics* **8**(1958), 459–466.

[90] H. Lebesgue, Sur la divergence et la convergence non-uniforme des séries de Fourier, *Comptes Rendus hebdomadaires des Séances de l'Académie des Sciences, Paris* **141**(1905), 875–877.

[91] H. Lebesgue, Sur les intégrales singulières, *Annales de Toulouse* **1**(3) (1909), 25–117.

[92] I. Leonard and J. Whitfield, A classical Banach space: ℓ_∞/c_0, *Rocky Mountain Journal of Mathematics* **13**(1983), 531–539.

[93] J. Lindenstrauss and L. Tzafriri, *Classical Banach Spaces*, Lecture Notes in Mathematics **338**, Springer-Verlag, New York, 1973.

[94] J. Lindenstrauss and L. Tzafriri, *Classical Banach Spaces I. Sequence Spaces*, Ergebnisse der Mathematik und ihrer Grenzgebiete bd. 92, Springer-Verlag, New York, 1977.

[95] J. Lindenstrauss and L. Tzafriri, *Classical Banach Spaces II. Function Spaces*, Ergebnisse der Mathematik und ihrer Grenzgebiete bd. 97, Springer-Verlag, New York, 1979.

[96] E. J. McShane, Linear functionals on certain Banach spaces, *Proceedings of the American Mathematical Society* **1**(1950), 402–408.

[97] D. Maharam, On homogeneous measure algebras, *Proceedings of the National Academy of Sciences USA* **28**(1942), 108–111.

[98] J. Marcinkiewicz, Sur l'interpolation d'opérations, *Comptes Rendus hebdomadaires des Séances de l'Académie des Sciences, Paris*, **208**(1939), 1272–1273.

[99] V. Mascioni, Topics in the theory of complemented subspaces in Banach spaces, *Expositiones Mathematicae* **7**(1989), 3–47.

[100] R. Megginson, *An Introduction to Banach Space Theory*, Graduate Texts in Mathematics 183, Springer-Verlag, New York, 1998.

[101] G. Metafune, On the space ℓ_∞/c_0, *Rocky Mountain Journal of Mathematics* **17**(1987), 583–587.

[102] P. Meyer-Nieberg, *Banach Lattices*, Springer-Verlag, New York, 1991.

[103] D. P. Milman, On some criteria for the regularity of spaces of the type (B), *Doklady Akademiia nauk SSSR* **20**(1938), 243–246.

[104] A. F. Monna, *Functional Analysis in Historical Perspective*, Wiley, New York, 1973.

[105] A. F. Monna, Hahn–Banach–Helly, *The Mathematical Intelligencer* **2**(4) (1980), 158.

[106] S-T. C. Moy, Characterization of conditional expectation as a transformation on function spaces, *Pacific Journal of Mathematics* **4**(1954), 47–63.

[107] M. Nakamura, Complete continuities of linear operators, *Proceedings of the Japan Academy* **27**(1951), 544–547.

[108] M. Nakamura and S. Kakutani, Banach limits and the Čech compactification of a countable discrete space, *Proceedings of the Imperial Academy, Tokyo* **19**(1943), 224–229.

[109] N. K. Nikolśkij, ed., *Functional Analysis I*, Encyclopedia of Mathematical Sciences, Vol. 19, Springer-Verlag, New York, 1992.

[110] W. Novinger, Mean convergence in L_p spaces, *Proceedings of the American Mathematical Society* **34**(1972), 627–628.

178   *References*

[111] W. Orlicz, Über unbedingte Konvergenz in Funktionräumen, *Studia Mathematica* 1(1930), 83–85.

[112] R. E. A. C. Paley, A remarkable series of orthogonal functions, *Proceedings of the London Mathematical Society* 34(1932), 241–264.

[113] A. Pełczyński, Projections in certain Banach spaces, *Studia Mathematica* 19(1960), 209–228.

[114] A. Pełczyński, A note on the paper of I. Singer "Basic sequences and reflexivity of Banach spaces," *Studia Mathematica* 21(1962), 371–374.

[115] A. Pełczyński, A proof of the Eberlein–Šmulian theorem by an application of basic sequences, *Bulletin de l'Academie Polonaise des Sciences* 21(9) (1964), 543–548.

[116] A. Pełczyński and I. Singer, On non-equivalent bases and conditional bases in Banach spaces, *Studia Mathematica* 25(1964), 5–25.

[117] B. J. Pettis, A note on regular Banach spaces, *Bulletin of the American Mathematical Society* 44(1938), 420–428.

[118] B. J. Pettis, A proof that every uniformly convex space is reflexive, *Duke Mathematical Journal* 5(1939), 249–253.

[119] R. S. Phillips, On linear transformations, *Transactions of the American Mathematical Society* 48(1940), 516–541.

[120] H. R. Pitt, A note on bilinear forms, *Journal of the London Mathematical Society* 11(1936), 171–174.

[121] H. Rademacher, Einige Sätze über Reihen von allgemeinen Orthogonalfunktionen, *Mathematische Annalen* 87(1922), 111–138.

[122] W. O. Ray, *Real Analysis*, Prentice-Hall, Englewood Cliffs, N.J. 1988.

[123] F. Riesz, *Oszegyujtott Munkai* (Collected Works), 2 vols., Akademiai Kiado, 1960.

[124] H. P. Rosenthal, The Banach spaces $C(K)$ and $L^p(\mu)$, *Bulletin of the American Mathematical Society* 81(1975), 763–781.

[125] H. P. Rosenthal, Some recent discoveries in the isomorphic theory of Banach spaces, *Bulletin of the American Mathematical Society* 84(1978), 803–831.

[126] H. P. Rosenthal, The unconditional basic sequence problem, in *Geometry of Normed Linear Spaces*, R. G. Bartle, N. T. Peck, A. L. Peressini, and J. J. Uhl, eds., Contemporary Mathematics 52, American Mathematical Society, 1986.

[127] H. P. Rosenthal, Some aspects of the subspace structure of infinite dimensional Banach spaces, in *Approximation Theory and Functional Analysis*, C. Chui, ed., Academic Press, Boston 1991.

[128] H. L. Royden, *Real Analysis*, 3rd ed., Macmillan, New York 1988.

[129] W. Rudin, *Functional Analysis*, 2nd ed., McGraw-Hill, New York 1991.

[130] H. H. Schaefer, *Banach Lattices and Positive Operators*, Grundlehren der mathematischen Wissenschaften bd. 215, Springer-Verlag, New York, 1974.

[131] J. Schauder, Zur Theorie stetiger Abbildungen in Funktionalraumen, *Mathematische Zeitshcrift* 26(1927), 47–65.

[132] J. Schauder, Eine Eigenschaft des Haarschen Orthogonalsystems, *Mathematische Zeitshcrift* 28(1928), 317–320.

[133] J. Schur, Über lineare Transformationen in der Theorie der unendlichen Reihen, *Journal für die reine und angewandte Mathematik* 151(1921), 79–111.

[134] C. Swartz, The evolution of the uniform boundedness principle, *Mathematical Chronicle* 19(1990), 1–18.

[135] C. Swartz, An addendum to "The evolution of the uniform boundedness principle," *Mathematical Chronicle* **20**(1991), 157–159.

[136] A. Shields, Years Ago, *The Mathematical Intelligencer* **9**(2) (1987), 61–63.

[137] W. Sierpinski, *Cardinal and Ordinal Numbers*, Monografie matematyczne, t. 34, 2d ed., Warszawa [PWN], 1965.

[138] G. F. Simmons, *Introduction to Topology and Modern Analysis*, reprint, Krieger, Malabar, FL 1986.

[139] A. Sobczyk, A projection of the space (m) onto its subspace (c_0), *Bulletin of the American Mathematical Society* **47**(1941), 938–947.

[140] M. H. Stone, Applications of the theory of Boolean rings to general topology, *Transactions of the American Mathematical Society* **41**(1937), 375–481.

[141] S. Szarek, On the best constants in the Khinchin inequality, *Studia Mathematica* **58**(1976), 197–208.

[142] W. A. Veech, Short proof of Sobczyk's theorem, *Proceedings of the American Mathematical Society* **28**(1971), 627–628.

[143] R. Whitley, Projecting m onto c_0, *The American Mathematical Monthly* **73**(1966), 285–286.

[144] R. Whitley, An elementary proof of the Eberlein–Šmulian theorem, *Mathematische Annalen* **172**(1967), 116–118.

[145] A. Wilansky, *Topology for Analysis*, Xerox College Publishing, Lexington, Massachusetts, 1970.

[146] S. Willard, *General Topology*, Addison-Wesley, Reading, MA 1970.

[147] P. Wojtaszczyk, *Banach Spaces for Analysts*, Cambridge Studies in Advanced Mathematics 25, Cambridge University Press, New York, 1991.

[148] K. Yosida, *Functional Analysis*, Die Grundlehren der mathematischen Wissenschaften bd. 123, Springer-Verlag, New York, 1965.

[149] K. Yosida and E. Hewitt, Finitely additive measures, *Transactions of the American Mathematical Society* **72**(1952), 46–66.

[150] A. Zygmund, *Trigonometric Series*, 2 vols., 2nd ed., reprinted, Cambridge University Press, New York, 1979.

[151] A. Zygmund, Stanislaw Saks, 1897–1942, *The Mathematical Intelligencer* **9**(1) (1987), 36–43.

Index

vector lattice, 154
Vitali–Hahn–Saks theorem, 140

Walsh functions, 86
weak convergence, 42, 43, 56, 116, 123, 138, 139, 145
weak topology, 133, 169, 170
weak* basis, 72

weak* compact sets, 133, 147
weak* convergence, 72, 123
weak* topology, 133, 134, 147
weakly compact operator, 142, 143, 144, 147
weakly compact sets, 133, 138, 140, 147
weakly null sequence, 42
weakly sequentially complete, 140
Weierstrass theorem, 129